CHASING PLANTS

Journeys with a botanist
through rainforests, swamps,
and mountains

CHASING
PLANTS

CHRIS THOROGOOD

Kew Publishing
Royal Botanic Gardens, Kew

The University of Chicago Press

© The Board of Trustees of the Royal Botanic Gardens, Kew 2022
Text and images © Chris Thorogood

First published in 2022 by Royal Botanic Gardens, Kew, Richmond, Surrey, TW9 3AB, UK www.kew.org
and The University of Chicago Press, Chicago 60637, USA

Royal Botanic Gardens, Kew
ISBN: 978 1 84246 764 0
eISBN: 978 1 84246 768 8

The University of Chicago Press
ISBN-13: 978-0-226-82353-9 (cloth)
ISBN-13: 978-0-226-82421-5 (e-book)
DOI: https://doi.org/10.7208/chicago/9780226824215.001.0001

British Library Cataloguing in Publication Data
A catalogue record for this book is available from the British Library.

Library of Congress Cataloging-in-Publication Data

Names: Thorogood, Chris, author, illustrator.
Title: Chasing Plants : journeys with a botanist through rainforests, swamps, and mountains / Chris Thorogood.
Description: Chicago : The University of Chicago Press ; Royal Botanic Gardens, Kew, 2022. | Includes bibliographical references and index.
Identifiers: LCCN 2022001418 | ISBN 9780226823539 (cloth) | ISBN 9780226824215 (ebook)
Subjects: LCSH: Thorogood, Chris—Travel. | Plants.
Classification: LCC QK5 .T48 2022 | DDC 581—dc23/eng/20220126
LC record available at https://lccn.loc.gov/2022001418

Design: Ocky Murray
Typesetting and page layout: Nicola Thompson, Culver Design
Project manager: Lydia White
Production manager: Jo Pillai
Copy-editing: Michelle Payne
Proofreading: Catherine Bradley

Printed and bound in Italy by L.E.G.O. S. p. A.

For information or to purchase all Kew titles please visit shop.kew.org/kewbooksonline or email publishing@kew.org

Kew's mission is to understand and protect plants and fungi, for the wellbeing of people and the future of all life on Earth.

Kew receives approximately one third of its funding from Government through the Department for Environment, Food and Rural Affairs (Defra). All other funding needed to support Kew's vital work comes from members, foundations, donors and commercial activities, including book sales.

CONTENTS

PREFACE

JUST IMAGINE IT: your parents on their hands and knees groping at a swarm of crickets unleashed from an upturned box; your teenage sister screaming at toads spawning in the bath; squirting cucumbers launching a raid of missiles down the stairs; and the gut-wrenching stench of a freshly unfurled dragon arum wafting through the front door. This is what I subjected my family to.

I was entranced by living things as a kid. I was never happier than when I was planting a seed and watching it grow, prodding something in a rock pool or releasing a butterfly freshly emerged from its chrysalis in a jar. Science was my favourite subject at school. I'd look at a living thing and think, how does this work? As a teenager I worked at a marine aquarium and foraged among the rocks at low tide for unusual sea creatures. My bedroom was a jungle of jars, pots and tanks crammed with botanical curiosities. I'd document all these things carefully, painting and illustrating the plants I grew – seeking to make sense of them. I was destined to be a botanist.

TODAY MY WORK takes me all over the world: over deserts and up mountains; through forests and into swamps. I still look at a living thing and think how does this work? And along the way I store up colourful memories of the plants I see, which I later fix

*Nepenthes
burbidgeae*

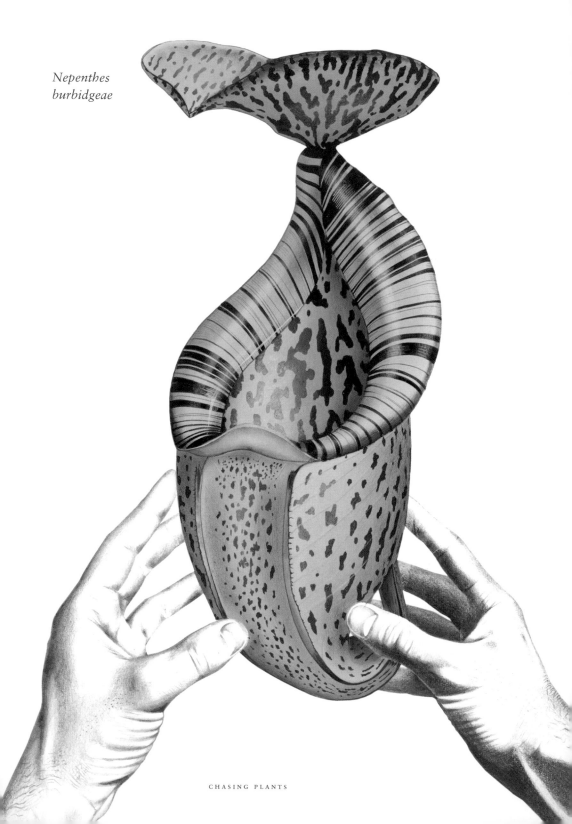

in paint. I've always painted the world around me. From time to time people tell me my illustrations remind them of Marianne North's. She was a botanical artist who painted her way around the plant world in the 19th century. Like most female artists throughout history, North did not receive the recognition she deserved during her lifetime. Today her 832 paintings of plants in their natural habitats crowd the walls of Kew Gardens' Marianne North Gallery, like a giant botanical jigsaw puzzle. I remember gazing up them as a boy, scanning the walls for pitcher plants. Decades later, I'd scan trees for them, just as she did; download my visions in oils, just as she did. And perhaps she was a little bit obsessed by plants, just as I am. I like to think that she was.

Many of the plants you will find in this book are the subjects of my scientific research today. One line of my work examines how carnivorous and parasitic plants evolved to look and behave the way they do. Carnivorous pitcher plants, which feature prominently in my paintings, produce traps derived from leaves, to attract, capture, kill and digest prey, to enable them to survive in nutrient-poor environments. In my twenties I spent time in Borneo, where I was struck by their bewildering array of shapes and sizes. Now, research shows that their assorted geometries mirror their diet. For example, the magnificent pitchers of *Nepenthes rajah* feed on manure: tree shrews scamper onto the pitchers and leave their nutrient-rich droppings behind in them. And that is why its pitchers are so sturdy: they are animal toilets.[1]

What can we learn from nature? Living things have evolved exquisite solutions to cope with the challenges they face, and these can inspire design in technology. Water-repellent lotus leaves, water-collecting wing-cases of desert beetles and water-removing gecko skin are some of the many organisms that have solved technological challenges relating to water movement. I work with physicists to explore potential plant-based solutions. Take the rim of the carnivorous pitcher plant. When wet, the rim becomes slippery, which leads insects to slide off it, along a series of chutes, into the trap. Creating artificial surfaces inspired by pitcher plants, we revealed a potential mechanism for droplet transport, guided by chutes. We found they trap, retain and direct the travel of droplets with absolute precision – just as they guide insects into the traps in nature.[2] Such a system could be a means of transporting and sorting droplets along pre-determined pathways in artificial devices such as ink-jet printers.

What can plants do for us? I also work with scientists around the world to understand the diversity of desert hyacinths (*Cistanche*). These curiously beautiful plants, which feature prominently in this book, may form part of a solution to the global problem of desertification (land degradation). Desert hyacinths are parasitic on the roots of desert shrubs including saxaul and tamarisk. Both these shrubs can be planted to form stabilising 'shelter forests' to halt desertification, which is fast emerging as a global crisis. In China, where desert hyacinths are prized for food and herbal medicine, farmers have started to grow them alongside these shelter forests as an ancillary crop. If we can grow desert hyacinths on a global scale, maybe we can achieve two goals at once: meeting people's requirements for food and medicine, while reducing the need to harvest rare wild desert hyacinths under threat.[3] But first we need to make sense of their diversity so we understand which ones to protect and which ones to cultivate. This is where taxonomy – the discipline that helps scientists understand and organise the dazzling diversity of life on our planet – comes in.

We can only protect what we know to exist. That's why botanists are in a race against time to find and describe new species so they may be conserved. Recently, I worked with local botanists in Malaysia to describe a new species that, miraculously, grows along a popular tourist track running through a mountain forest.[4] It has gone completely overlooked. But let's be clear: many species 'discovered' around the world every year have for millennia been known and used by people who go unrecognised.

Desert hyacinths (*Cistanche deserticola*) may form part of the solution to the global problem of land degradation.

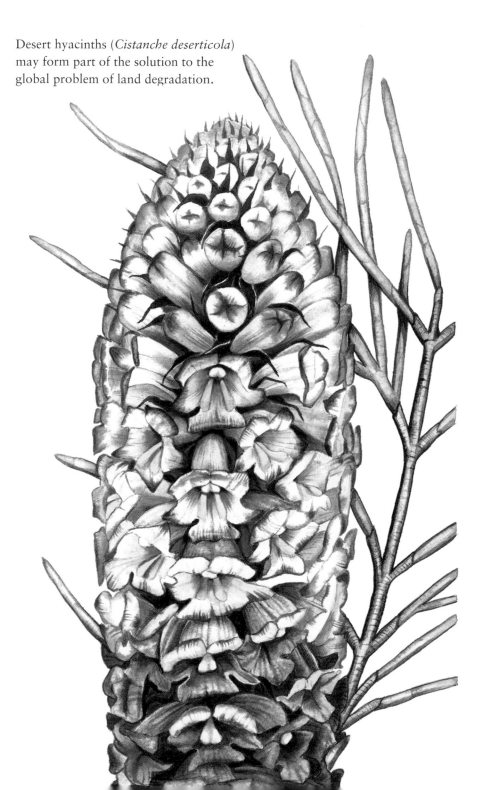

This imbalanced Western legacy of discovery needs to change. One way that botanists can bring about this change is to work *with* local people around the world, exploring its diversity and protecting it *together*.

THE JOURNEY I take us on around the world in this book is not chronological and I've deliberately omitted some dates to save you from getting dizzy – the clifftops and typhoons will do that well enough on their own. The book has its diversions; it hops about all over the place, just as I have. In writing it, I dipped in and out of my field notes, just as I imagine you, the reader, might dip in and out of the book. The passages in 'To Pitcher Plant Paradise' are from a diary I kept in Borneo in my early twenties; the snatches of time spent in Japan are much more recent. The purpose of each trip was different: as you will see, in Japan, I carried out conservation work – collecting plants for seed-banking and carrying out vegetation surveys. In the Canary Islands I worked with local botanists and ecologists, documenting the flora and planting seedlings alongside the local community. My adventures across the Levant – some with other botanists, others forged alone – were for research, both for taxonomic purposes and for field guides to the region's flora. My time spent with pitcher plants was more selfish: it fulfilled my childhood dreams about plants, the seeds of which, you will discover, were sown long ago outside IKEA (surprisingly). That's why I finish my journey on Mount Kinabalu in Borneo – the place I lay awake thinking about as a child: it is as much a beginning as an ending.

Why have I turned my diaries into a book? Regardless of when the trips took place and why, they are united by an uncontainable passion for plants that I need to share. 'You'll need to be clear that you're not just some isolated nerd looking for plants, it *must* have purpose!' pleads my publisher. Clearly I *am* a plant nerd, but she's right: botanists like me have a vital role to play in raising awareness of plants. We rely on them for our very existence: for food, clothes and medicines – and, as we're discovering more and more, for our mental health and well-being. They energise our planet. Yes, we've never needed plants more than we do now. But far more than this, plants have an intrinsic value of their own. We share the biosphere – this thin layer of life we call home – with hundreds of thousands of plant species that existed long before us, and we have a duty of care to protect them. Yet two in five are now threatened with extinction.[5] They're losing the fight against the threats they face amid a growing human population; some disappear before we even know they exist. Worse still, their plight goes largely unnoticed – part of a problem referred to metaphorically as Plant Blindness. Put simply, we don't even notice them.

What can we do about it? Perhaps we can bring plants out of the shadows by portraying them differently: by showing their intrigue and their character, beyond their being a beautiful backdrop for animals to exist against. We can explain why we must protect them just as we must protect animals, and challenge the perception of what botanists do and why we care so deeply. I hope that my book will do this in some small way. Perhaps it will intrigue someone, maybe a student – the kind of person who looks at a living thing and thinks how does this work? – to dream of becoming a botanist one day. Perhaps, in turn, they might explore, wonder and seek to protect what they have read and dreamed of, and leave this world a little better than they found it.

THE WORLD

Key locations visited in the book are labelled below in CAPITALS.

North
America

BRITAIN
&
IRELAND

THE
CANARY
ISLANDS

N
W E
S

South
America

urope

MACEDONIA

Asia

JAPAN

THE
CRETE MIDDLE
CYPRUS EAST

Africa

BORNEO

SOUTH
AFRICA

Australia
& Oceania

BEFORE WE GO

EVER BEEN OBSESSED with something? I mean *really* obsessed? Ever lain awake dreaming about it: a new car perhaps, your dream house, a person? I guess we all have them, only my obsession is a little offbeat. Mine is *plants*.

It has always been. I've chased plants for as long as I can remember: I've learned the language of them, how to read the land with them and how to speak to its people through them. Knowing plants lets you see a place in differently, read the forest's mind, and listen to the mountain.

Once you've thought of something long and hard enough, you can't unthink it. Right? Once you've hauled yourself up a wet mountainside in pursuit of a pitcher plant, macheted your way to that orchid, or divined the world's largest flower in your mind's eye, you have to do it for real, don't you? Find them, whatever it takes. Get to know them, their haunts; bathe in their beauty.

Well I've courted danger in pursuit of plants, crossed barbed wire to find them, gone over the cliff to admire them, fallen for them and stared into the barrel of a gun. And I've never felt so alive. Along the way, I've learned that you don't *need* to do any of this. For as every plant enthusiast knows, it's fulfilling to track down a striking work of nature wherever it may be – the woods, the golf course rough, even the retail park. And when you do, there's no greater feeling on earth: it's electrifying.

I hold onto that electricity, and later, I transform it into colour. I conjure up my plants onto canvas, steadily bringing back to life my glorious moments with them by brush, collaging experience with imagination in my paintings. Together with my field diaries, they tell a story, one of a boy who dreamed of becoming a botanist.

Now let me share with you how I've scribbled and sketched my way around the world fulfilling that dream.

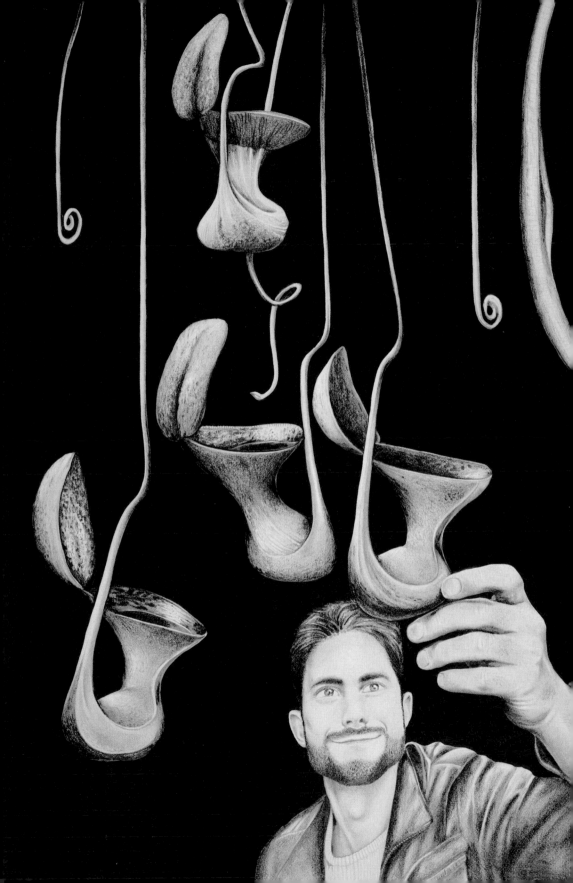

1

ON A TRIP TO IKEA

BRITAIN AND IRELAND

ON A TRIP TO IKEA

BRITAIN AND IRELAND

Expect extraordinary plants in ordinary places.

MY JOURNEY BEGINS in a retail park in Essex, south-east England. Bear with me – I promise this journey will become more exotic. But it was not on a mountain nor in a rainforest that I found the plant that was to change my life. No, no. It was outside IKEA among the coke cans. Here I found my first broomrape: a leafless ghoul that snatched at the roots of other plants, unseen. It had no chlorophyll – not an ounce of it – having forsaken the wholesome photosynthetic existence of its ancestors for a life of crime, stealing strength from its victims. This botanical enigma ignited my curiosity as a child, challenging my notion of how a plant behaved, and even what a plant was. It became my compass.

On my grandmother's shelf sat a cherished row of the Victorian illustrator Anne Pratt's six-volume work describing the flora of Great Britain. When I wasn't training toads in the garden, setting off squirting cucumbers or ruining a new carpet with paint, I'd spend time poring over these books. One of the musty volumes, describing a tribe of plants 'parasitic on the roots of other vegetables', became well-thumbed. Before I knew much about plants, their tawny, leafless forms leaped off the page at me. 'Many on first looking at it,' explained Anne, 'have believed it to be the remains of a flower from which the summer's sun had withered away all the beauty.'[6] Not me. This group of plants had a curious beauty of its own – and I wanted to find one more than anything.

It happened on a family shopping trip in my teens. 'Stop!' I shouted from the car, pointing to a bank of shrubbery in the car park. 'Not again!' chorused my family as we screeched to a halt. I jumped out to examine it and collect its seeds. My first one. Actually, a whole forest of them: common broomrapes (*Orobanche minor*) growing in their hundreds, on the roots of *Brachyglottis*, an ordinary-looking ornamental shrub. 'Mr Loudon,' continued Anne, referring to the Scottish botanist and garden designer 'says

Orobanche minor var. *heliophila* growing in a retail park in Essex, south-east England.

that any of the broomrapes may be made to grow in the garden on furze and broom'. I fancied that I might cultivate them for myself in my parents' back garden. Reading that common broomrape was partial to the roots of red clovers, this was the host I selected for my garden trial. Only to my disappointment, a year later, my crop failed: not one broomrape emerged. The following year, I tried the experiment again, this time using a potted *Brachyglottis* shrub – the plant I'd found the broomrape feasting on at IKEA. This time, to my delight, dozens of broomrapes pushed their way up out of the tub like bunches of purple asparagus. Mr Loudon was right about one thing: broomrapes *can* be grown in the garden; but as to their choice of host, they are far fussier than he had supposed.

Keeping the results of my early garden experiments steadily in view, I made these curiously illicit parasitic plants the focus of my early career in research. My PhD probed deeper into the secret lives of broomrapes. I set out to understand whether common broomrape was as catholic as the books claimed regarding its choice of host. And so it was – clifftop dangling and golf ball ducking – that my vertiginous quest began. I searched for them all over the place, crawling under fences, over clifftops and appearing in people's back gardens. An army of enthusiasts mobilised by the Botanical Society of the British Isles sent me samples by post from up and down the country. My mother was persuaded to collect specimens from a churchyard and got caught, trowel in hand, by the curious vicar. We were all knee-deep in broomrapes.

Back in the lab, I grew them in petri dishes, scrutinising their intricacies: their DNA, which plants they would feed from, which they wouldn't. And I found that seemingly identical populations behave very differently indeed. In fact, they are evolving new species right under our noses. This was my discovery: isolated from one another, broomrapes were silently leading parallel lives in their various haunts, from clifftop turf to golf course roughs, even in IKEA car parks.

Orobanche minor var. *pseudoamethystea* growing by a golf course in Kent, south-east England.

KENT

I arrive at Dover Priory train station in Kent, on the hottest day of the year, in July. The sea air is filled with the shrill siren of acrobatic gulls and a sharp, sweet smell like salt and vinegar. The seaside town comprises a port and a ribbon of houses nestled at the foot of a pleated white cliff, all presided over by a great grey castle. It has a bustling, purposeful feel, a place where people look distracted. After dumping my bags at the local bed and breakfast, I make my way eagerly towards my destination: the stretch of coastline between Dover and Deal, along which grows one of Britain's very rarest plants.

A few miles from Dover lies the village of Saint Margaret's at Cliffe. Once a well-to-do holiday resort, it feels like it might not have changed a great deal since the 1940s. A steep, green lane pivots its way awkwardly around twisted pines and compact shrubbery mounded by coastal gales, with spiky cordylines poking out. I catch glimpses of manicured lawns and shady dining room windows. Eventually, the narrow lane flattens and peters out into the bay, where a tangle of wild ivy and clematis bleed into a straw-parched lawn.

The grey sea gently slaps and rakes the orange and grey shingle. The air tastes salty. I crunch my way past suncream-streaked holidaymakers draped along the beach like seals. A woman leans against the breakwater, adjusting a purple flip-flop; another nearby is shaking a towel. Both are as pink as prawns. I reach a quartet of white villas at the far end of the beach. One of these, I read, was a holiday

A 'mermaid's purse', the egg case of a lesser spotted catshark.

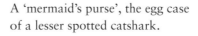

home called 'White Cliffs' owned by the 1940s playwright Noel Coward. Beyond the villas, the bathers vanish and the beach becomes wild. Here the sea has thrown up a curious assortment of flotsam and jetsam including a bundle of creamy mermaids' purses: the fancifully-named leathery egg cases of the small catshark, along with some chalky cuttlebones. Their curious, shapely forms belong to another world – a hard-cut watery place where creatures go about their business unseen. I pick my way through a messy web of intense-smelling crispy seaweed. Just beyond it I spot a mound of chalky rubble, slumped at the foot of the cliff. This looks promising.

The sun is searing now and a heat haze shimmers on the horizon. I could be in Spain. I scramble up the steep, loose scarp. The constant crash and hiss of the sea seems louder by the cliff. A few metres up, my palms white and dusty, I find a succulent mass of grey-leaved rock samphire sprawling over the rocks. I nibble on a juicy leaf that has a unique salty, spicy taste – with a hint of the smell of furniture polish. Every plant has its own unique fragrance and taste, I've discovered, and I like to get to know them (unless they are poisonous). I've learned to tell apart identical-looking species of ivy by their smell alone.

With glee I spot the first broomrape of the day among the patchy vegetation: a fat spike of carrot broomrape (*Orobanche minor* subsp. *maritima*) pointing to the sky. Not the plant I'm after, but an exciting find nonetheless. Carrot broomrape was first recorded in Britain in 1845 by the Reverend W. S. Hore from on the cliffs at Whitsand Bay in Cornwall. Since then it has baffled botanists with its superficial similarity to other closely related broomrapes. But its affinity for sea carrots (upon which it feeds),

Orobanche picridis growing on the White Cliffs of Dover, Kent, south-east England.

its dark purple stems and the two yellow knots on the lower lip of the flowers are quite distinctive. I bag a sample for a future DNA extraction back in the lab, and make my way back up to the top of the cliffs, the taste of samphire lingering.

'Half-way down, hangs one that gathers samphire; dreadful trade!' writes Shakespeare in *King Lear* (Act 4, Scene 6), referring to the dangerous pursuit of gathering the cliff-dwelling rock samphire. But my quest is far more dangerous than a samphire harvest, I discover: picris broomrape (*O. picridis*) is what I'm after, and this predominantly Mediterranean species is a cliffhanger. Britain is the northern limit of its range, where it's virtually confined to the White Cliffs of Dover and the Isle of Wight. Britain's rarest native broomrape, this thermophile likes the dry rocky ledges of cliffs that bake in the sun, akin to warm Mediterranean scrub.

Scouring the crumbly edge of the cliffs between Dover and Deal I find a dozen of the pearl-coloured spikes sprouting among the thin grass scorched dry by the hot Kentish sun. I crouch down and examine their features closely. 'Good' specimens have blackish stigmas, long, twisted bracts and hairy filaments; there's no mistaking these. Smaller ones are trickier because they overlap considerably with their more pedestrian cousin, the common broomrape (*O. minor*) that also grows here. I take a little tissue from everything I find. But for the DNA I plan to examine back in the lab, I need more samples: there's no avoiding it, I need to go over the edge of the cliff.

The late afternoon sun is still hot. Blue butterflies flutter aimlessly over the clifftop turf to an orchestra of burring crickets. The milky-blue channel glistens. A little meandering path snakes in and out of the briers, tussocky paddocks with horses and clumps of stinking iris (*Iris foetidissima*); I amuse myself by crushing a leaf, the scent of which is just like roast beef. Soon I see the busy Port of Dover stretching out onto the horizon to my left. It looks like an industrial estate of cars, cranes and tubes, sinking into the sea. I peer over the great chalky precipice. A long way down, frothy seawater is silently plucking the bouldery foot of the cliff. Dissolved chalk plumes out into the depths, in slow motion, like milky clouds pouring into ink. And there, a few metres below on a rocky ledge, is exactly what I'm looking for: an irresistible clump of ivory-coloured spikes of picris broomrape.

I'm not a qualified climber. I don't speak the lingo, do the belaying, tie the knots – none of that. But, in a way, I've trained my whole life by looking for plants. I inch my way down, my back against the cliff, with clammy palms. *Don't look down*. Slowly, I edge my way to the shelf. It's only a foot wide and rather less stable than I'd hoped. However, I can sit on it without too much difficulty, with my legs dangling over the edge. For a magical moment, I'm alone with these special plants, the White Cliffs of Dover above, below and behind me and the Channel stretching out in front of me. No one can even see I'm here – probably just as well because they'd surely call the fire brigade if they could.

As if reading my mind, a siren startles me. I realise it's the ferry terminal below. 'May I have your attention please. This is a security announcement . . .' reverberates a slow, female voice that echoes about the cliffs. It sounds uncannily like Judi Dench. The spell broken, I set to work, examining, measuring, collecting and scribbling. I twist myself carefully into position so that I may take a photograph or two. 'If you see something that doesn't look right . . .' continues Judi. Probably best to make an exit now in case I'm spotted from the ferry terminal below. I look up to scan the vertical cliff stretching up between me and the path. As I stand, a white stream of chalky debris dislodges noisily from under my feet and vanishes into the abyss. I grab the rock, my palms dripping with sweat. This is a risky business. I reassure myself that since I managed to find a route down here, I can surely get back up without too much difficulty. Out of nowhere, an enormous white gull looms up from underneath me, screaming piercingly. To my horror it tries to extricate me from the cliff! Like something out of a nightmare, the griffon wheels around me repeatedly, lunging with stabbing orange feet extended like some hideous great raptor zeroing in on its prey. Is this normal behaviour for a seagull? 'Get *off* me, you *bastard*,' I shout, wobbling, and nearly lose my footing. Panicking, I grab a grassy tussock for balance. I scrabble my way up, flinging further flurries of rock. One careful step at a time, I manage to grasp the rock, pulling myself up to the top of the cliff. The gull laughs loudly.

This is extreme botany.

Lightheaded, I haul myself back to safety, my palms a mess of chalk, sweat and blood. The gull seems satisfied that I'm no longer trespassing in its domain and circles sulkily away. I dust myself down and head quietly back to the bed and breakfast, telling myself I'll never do that again. But I do.

MONMOUTHSHIRE

I arrive in South Wales on a soaking wet day in June. The purpose of my visit is to look for a rare form of broomrape (*O. minor*) that has yellow spikes of flowers, crowded atop stout, scaly stems. Examined carefully by only a few botanists, the form I'm after may grow only here, around the docks, and I have a delicious sense of anticipation at the prospect of finding a plant I've not seen before.

The lane towards the docks smells of warm, wet tarmac. It's flanked by high grey railings, holding back a forest of buddleia. Above me looms a web of pylons. I spot a woman wandering in the opposite direction, in a pink dressing gown and slippers; an interesting choice of clothing in a rainstorm. Somewhere in the distance, a motorbike revs and roars. I reach the end of the grey lane where a sign warns against fly-tipping next to a messy heap of cans, cigarette packets and various discarded plastic items; close by another sign says, 'Welcome to the port'.

Orobanche minor var. *lutea* growing by the docks in Monmouthshire, South Wales.

It is not altogether clear where I can and can't go, but the man at the entrance to the docks doesn't object to me exploring the area to my left, so I go in. The patchwork of grey, gritty spaces and sandbanks, dissected by a rusty railway track, is covered in an uneven carpet of straw-coloured grasses and ruderal weeds. I note at least 20 different plant species in under ten minutes. Everywhere I wander, the warm damp air is thick with a mélange of buddleia and creosote, all sweet and smoky. I splash my way through black puddles that have sprung up all over the place. After a short while I encounter my first stand of lemon-coloured broomrapes sprouting by the train track, pushing their way out of a sandy patch of rain-beaded grass and clover.

I get that familiar thrill of finding a plant for the first time, feel it coursing through my system like caffeine. The plants are small and immature, not ripe enough for close examination. I will need to find more. After a good hour of scouring the area, soaking wet and with tar fixed firmly in my nostrils, I head down a roadside track. Lorries flicker in and out of view through the soggy hedge. Then a flash of yellow catches my eye. At last, among silvery sun-bleached crisp packets and cigarette butts, I find a fat, mature broomrape specimen. Its thick, butter-yellow spike glows against the grit and inky weeds. Its flowers are congested and dome-like at the top, just as they should be. I note its location so that I can return to collect its seeds. This isn't the most salubrious of places to go plant hunting. But the thrill is just as good.

YORKSHIRE

'Are you alright, love?' asks a middle-aged woman with a broad Yorkshire accent, walking her terrier. She might well ask; I'm on my hands and knees, slithering into a hedge, a stone's throw from the main road. 'Oh I'm fine, thanks. I'm looking for a rare plant,' I say. She and her terrier stare at me blankly. 'I'm a botanist,' I explain. She looks at me suspiciously. I become conscious of the twig poking out of my hair and a bleeding scratch across my forehead. 'Right,' she says, nodding, looking unconvinced. 'Well then . . . good luck.' She wanders off, looking over her shoulder. She obviously thinks I'm up to something. I hope she doesn't call the police.

I can see why she's sceptical, though. This doesn't look like promising territory for a rare plant. Far from it. But it happens to be the best place to find a species that is very rare in the British Isles: thistle broomrape (*Orobanche reticulata*). So rare, in fact, that it evaded botanists for centuries. Today it grows at only a handful of sites on the Magnesian limestone of north-west Yorkshire, where it appears to be associated with historic Roman roads. Although today it seems more associated with motorways. Traffic murmurs in the background.

Back to the task in hand. I drag myself through the scratchy hawthorn and emerge in an open parcel of grass. I look around, picking the briers out of my clothes. To my sheer delight, thistle broomrapes are pushing up their muscular spikes all around me. Some are in full bloom, with thick, liver-coloured stems bristling with rose-tinted flowers. The largest come halfway up to my knee. I spend a good 30 minutes hopping from plant to plant, crouching, crawling on all fours and generally doing what we botanists do. And as the motorway purrs in the distance – the one that cuts right through their biggest remaining colony – I think about the precarious fate of this jewel in the British flora.

Orobanche alba growing out of the cliffs in County Antrim, Northern Ireland.

NORTHERN IRELAND

Today I'm accompanying the world's leading authority on historical families, William Bortrick, for a botanical excursion around Northern Ireland. As we zip around the coast in William's convertible, I scan the rugged landscape for unusual plants. William points out great landmarks from the car, including the sombre Garron Tower presiding over a great leafy plateau, and the historic Londonderry Arms in Carnlough, nestled in the lush Glens of Antrim – a hotel that Winston Churchill inherited from his great-grandmother, Frances Anne Vane, Marchioness of Londonderry. We stop at a small petrol station for a quick snack. Half an hour later we leave with a home-cooked lunch – you could stop at any house here and someone would happily cook you a meal. We park at a smallholding, the owner of which lives off the land quite well without electricity, and make our way wet-shod towards Northern Ireland's north-easterly corner, Fair Head.

Sinking into feather-soft domes of sphagnum moss, we crouch every now and again to examine a tiny plant at our feet. The marshy land rises gently up to the brow of a steep, grey cliff. Heather and bilberry form a patchwork, carved occasionally by a lime-green lush packed with cross-leaved heath (*Erica tetralix*) and fresh-looking ferns. Closer to the windy brink of the cliff little pink clouds of bell heather (*Erica cinerea*) appear between the rocky outcrops. Over the cliff edge spreads a softly forested slope, scribbled at the edges with black rock that falls softly into the sea. 'And that,' beams William, pointing to a grey ribbon of land under a milky sky, 'is Scotland.' We scan the horizon towards Ballycastle and Rathlin Island and peer down at vertical rocky columns like organ pipes where wild goats roam the grassy ledges. On the shelf of a wilfully bleak cliff gleam sky-blue flashes of sheep's bit (*Jasione montana*). Peregrines dive over us. Cows stare at us. And William – a man who examines people like I examine plants – asks about my childhood. I tell him of one spent with plants in IKEA car parks, and pet toads, as we weave the broken edges of this powerful grey landscape.

'YOU'LL STRUGGLE TO find them,' says the groundsman, rolling his cigarette paper.

'Yes, but Chris can sniff out a broomrape like a *truffle*!' explains William.

'You'll struggle to find those too,' the groundsman says, licking the paper. 'But Simeon will know where to look. He's a ranger for the National Trust, and there isn't a plant in Northern Ireland he doesn't know.'

We ask him how we might locate Simeon.

'You'll struggle to find him,' he mumbles, hand cupped around his cigarette. 'He's out on a run.'

We ponder this and decide perhaps it's best to set off, take our chances. Secretly, I'm sure I can intuit the plants' whereabouts if I'm in the area. *Think* like a broomrape.

But out on the road, as luck would have it, we see a pair of runners leaning against a car, panting.

'Which one of you is Simeon?' asks William.

'Why, what's he done?' asks one.

Just at that moment Simeon himself appears, and we quiz him on his knowledge of thyme broomrape (*Orobanche alba*). The groundsman was right, he knows precisely where the plant grows. We commit to memory his precise instructions with care, picking our way over wire fences, kissing gates and into gorse thickets in our minds' eyes. We thank him for his help and head east for the Causeway Coast. Forty metres east, to be precise, because just after we set off I shout, '*There!*' A stone's throw from where we've discussed thyme broomrape's rather exacting haunts, I spot the very plant sprouting out of the roadside cliff. We abandon the car in the road, to the consternation of the other drivers ('they won't mind, and it is an emergency of sorts,' says William, with a sweeping gesture). We stand on the roadside and scan the imposing cliff above us.

To know a place – know it properly – you have to immerse yourself in it, don't you? Grab it with both hands, feel its dirt beneath your fingernails. So that's what I do. I find the indentations for finger grips and haul myself up, one ledge at a time. I examine the little rockbound patches of thyme spilling over the crags, and inch my way up, dislodging flurries of basalt. Sweat dribbles down my temples. I shout to William below that the plants are in my sights now. And suddenly I'm surrounded by them: ruby-coloured broomrapes conjured out of the rock by long days of watery sun and plumps of rain sent from the Atlantic. Their chunky little spikes jut out in clumps – some a dozen strong, with still more pushing through the rock. Their stems are the colour and texture of overripe peaches and their shapely flowers smell faintly of cloves. I spend a precious moment with them; then the growing tailback on the road below brings me back down to earth.

A short distance from broomrape dreamland lies the Mussenden Temple. The little domed Grecian temple is perched high above the Atlantic in front of a rose-coloured, painterly sky. William (who holds a key to every historic building in Northern Ireland!) fumbles with the locks. 'Will you let us know when you're in so we can join yous for a wee glass of wine?' sings a friendly, heavily accented voice from a visitor on the path below; we all laugh at the ridiculousness of the situation. But three locks times six keys later we're in. A magical space opens in front of me – one that echoes cold marble, damp plaster and time. I learn that the temple was constructed in 1785 and formed part of the estate of Frederick, 4th Earl of Bristol, who served as the Church of Ireland Lord Bishop of Derry. I peer through the salt-misted windows at the rolling Atlantic that whispers through the glass. Back on the cliff the hunt for broomrapes continues against a coral sunset mist dripping into the sea.

Kelp on the breeze; moss beneath my feet; embers in the rock and embers in the sky. This was my day here.

SITE NOT DISCLOSED
(CLASSIFIED INFORMATION)

The things I'll do for a broomrape. The four of us slip into hi-vis vests, unfold our safety specs, clip hard hats into position and tie up our steel cap shoes beneath a pyramid of coal. It's not how I usually dress when looking for wildflowers. We look as though we're about to go down a mine. Fellow broomrape addict Fred Rumsey and I are here to verify a mysterious plant noticed for the first time recently by a bird ecologist called Darryl, who joins us along with our site host, Sophie. I can't say precisely where I am because this must be kept secret – we're on a private industrial estate and the visit has taken a month of paperwork to prepare. What I can say is that it's about the least likely place you'd expect to see a rare plant.

We inspect the map, planning our route. As with all good treasure hunts, we trace the Xs on the map that Darryl has scrawled for us. Sophie asks what time we plan to leave and we

Orobanche picridis growing
on a private industrial estate.

agree we'll need a good three hours here, given the size of the site. 'My wife'll kill me . . .' mutters Fred into his face covering. Decked in luminous orange PPE, we clomp along a road of graphite. It smells of hot asphalt. We pass a complicated web of industrial pipes large enough for a car to drive through, plumes of vapour, cranes and factory chimneys. Enormous, beeping, tank-like vehicles crunch along the grit, besprinkling the dust along the verges with water, leaving a stream of treacle in their wake. Everything here is black.

'*Please* be careful!' pleads Sophie. She could be forgiven if she is finding this a bit tiresome, as we keep darting off, leaving our hard hats in thickets and generally breaking several rules at a time – but really, bringing two botanists to such a place is asking for trouble. Fred crouches to observe some plant or other, wielding a bulky SLR camera oriented towards the road like a furtive speed camera operator lurking in the grass. The ash-black mounds along the road are flecked with blackened plants – lanky hawkweed oxtongues, rosettes of plantain, yarrow fronds – that sort of thing. Weeds are often defined as plants in the wrong place, aren't they? Speaking of those, something else is growing here too; something unexpected.

'There's one!' shouts Fred, happily. 'And another,' I chime in. Most are golden and crispy, but a handful are still clothed in fresh white flowers at the tops of their stems like teeth, all peppered with ash. That plant I courted danger with on the White Cliffs of Dover? The one I said was rare? Re-enter picris broomrape (*Orobanche picridis*)! Hundreds of them – sprouting from every bank, siding and mound of gravel. They're even growing out of the road. Against all the odds, we have discovered the largest population of one of Britain's rarest plants growing in the heart of a sprawling industrial estate. We can hardly contain ourselves, hopping from one mound to another, scurrying up spoil heaps, sliding down ramps, parting tufts of grass and flinging ourselves into verges. We attract a small audience of workmen, who wander over to see what all the fuss is about. Excitedly we show them the plant and explain why it's so special. They peer at the browning spikes poking out of the slag. They don't look convinced. 'Well, that's always been here!' says one. 'What about this one?' says another, pointing to a dandelion. 'Are they worth anything?' says a third. Soon half the shift is wandering along the road gawking at them. Everyone is knee-deep in broomrapes.

2

VAMPIRE HUNTING

SOUTH AFRICA

2

VAMPIRE HUNTING

SOUTH AFRICA

Namibia

Botswana

South Africa

Cape Town

Little Karoo Desert

Indian Ocean

Atlantic Ocean

*A line of ants leads me to them: a coven of gaping,
red-mouthed vampires.*

I WAS HOOKED on them. Parasitic plants – sap-sucking vegetable
vampires – they had their teeth in me. Their unique biology and
their curious forms grabbed me. But there was one in particular
that I had in my sights: *Hydnora,* dubbed 'the strangest plant
in the world'. Like the broomrapes, it has foregone leaves and
succumbed to a pilfering existence, feeding from the roots of
shrubs and trees, only emerging above ground to flower. You
know about it when it does! Its extraordinary, fist-like flowers
punch their way up and nothing gets in their way; they can even
burst through pavements and damage infrastructure. They open
to reveal a mouthful of white fangs (this *is* a vampire). These
so-called 'bait bodies' emit evil breath to attract inquisitive dung
beetles. Do you think I'm playing with the boundaries of non-
fiction here? I'm not. Nothing, repeat nothing, is stranger than
this plant.

Just a handful of *Hydnora* species occur across the semi-arid
regions of Africa, Madagascar and southern Arabia. All of them
bloom unpredictably, often in remote regions, and they can be
tricky to find. But I have a gift for sniffing out rare and unusual
plants. While spending a summer exploring South Africa's
Cape Floristic Region, this plant would be my next fix: I *had* to
find it.

THE ROAD FROM Cape Town meanders through vast rocky grey and orange canyons draped in fynbos (shrubbery). It smells like a herb garden. After a couple of hours, fynbos gives way to the Succulent Karoo Biome, an undulating semi-desert plain blanketed in grey, mounded bushes and meaty-looking euphorbias. I park at the Karoo Desert National Botanical Garden, at the foot of the Hex River Mountain range. The garden, I discover, is a succulent wonderland framed by blue mountain vistas on all sides. I chat to a friendly member of the garden's staff over a coffee about the paperwork necessary for seed collecting. Then we wander along the stony paths that meander around the rocky terraces of quiver trees and talk about plants.

After a happy half hour in the garden I set off into the rocky wilderness shimmering beneath the midday sun. The air smells herby like cannabis, mingled with the coconut-like scent of suncream that I'm applying hourly because the sun is so strong. All is quiet and I feel at one with the Karoo. My desert bathing is interrupted by a lonely angulate tortoise, knocking its way clumsily through the rocky veld. It seems unperturbed by my presence, and I watch it clamber awkwardly into the thicket.

All is quiet again. I search for a good hour among the scratchy bushes. I know just where to look: beneath the grey-green stands of *Euphorbia mauritanica* which jut out from among the boulders. I survey the landscape judiciously, inspecting each euphorbia one by one, peering into their pliable, waxy branches. Nothing. I lick the scratches that now criss-cross my arms; they taste of salt and metal. There is no path, and it becomes hard to keep track of where I am. I accidentally check the same euphorbia twice – they all look the same. The summer sun is fierce and it would be a bad idea to get lost in the desert with so little food and water. But I can sense that I'm close.

Hydnora africana in the Little Karoo Desert, South Africa.

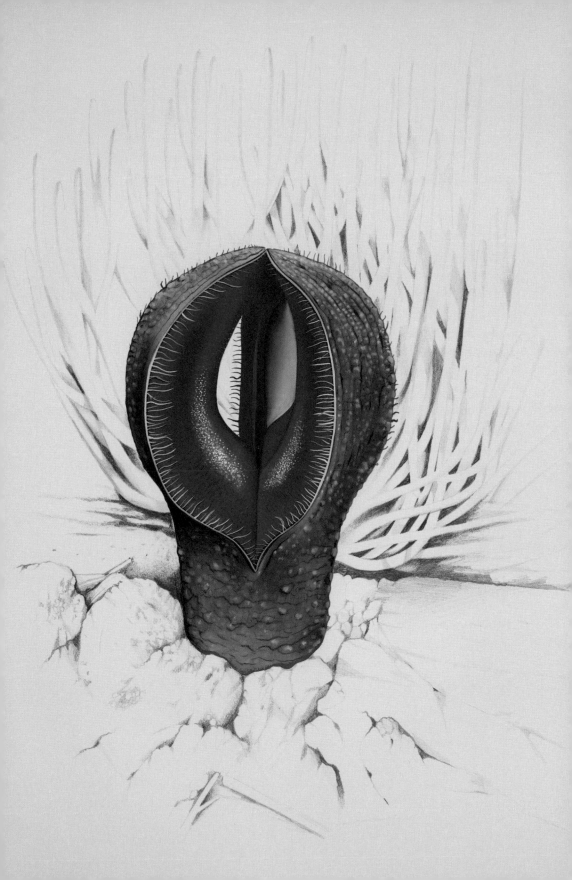

A watch line of large ants march purposefully across the desert. I follow them. The ants veer knowingly to the right, leading me to a fresh patch of euphorbias. And suddenly, there: *there* they are! A coven of red-mouthed vampires gape at me. At last. Giddy from sun and excitement, my temples pulse. I can't believe my eyes. A dozen are pushing their way out of the ground, each at various stages of their development. Most have withered, for it's late in the season, but some are still closemouthed, in tight bud. I flit from one to the next, then back again, unable to contain myself. I sketch a couple, peering into their curious red structures. One is in fruit. This is what I need to collect the seeds that I have permission to store back at the botanical garden. I dig it up as you would a large bulb and it lifts easily from the rocky red earth. It's the size of a small football, the colour of brown paper, and warty as a toad. I make an incision to check that it's ripe. Its interior is like a melon, with glistening pink and white pulp, streaked with ribbons of tiny seeds. I pause for a few moments to savour my prize.

I'm not sure I can surpass this feeling of exhilaration, alone in the depths of a desert with these striking works of nature; not sure I should even try to. But that's not how addictions work, is it? To get a hit like this again I'll need to find something wilder, more remarkable. Something *big*.

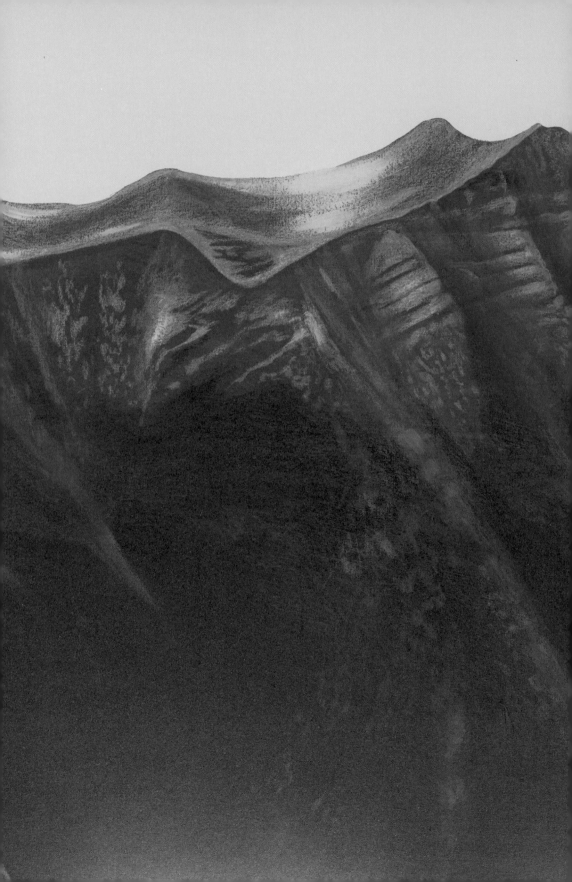

3

DRAGON SLAYING

MACEDONIA, CRETE AND CYPRUS

3

DRAGON SLAYING

MACEDONIA, CRETE AND CYPRUS

*Down, down, down I go, tumbling into
a thunder of dragons.*

EUROPE NUDGES ASIA in a commotion of mountains, gorges, steppes, plains and deserts. The region known as the Levant (the eastern Mediterranean, in its broadest sense) has a complex geology and climate. Here, cold northerly winds from continental Europe and Turkey smack warm breezes lent by the Sahara. A confusion of winds, seas and continents has hammered out a diverse flora over geological time; one coaxed out of the sun-dazzled rock and matured by various civilisations including the Egyptians, Greeks and Phoenicians. This is a land as rich in plants as mythology.

Botanical science is rooted firmly in the eastern Mediterranean. I tap these words into a laptop overtopped by a tower of ten weighty volumes in an Oxford library, torched by diagonal shards of dusty light streaming through lofty windows. They belong to the *Flora Graeca*, considered one of the finest floras ever produced. Written by the professor of botany John Sibthorp and illustrated by the young Austrian Ferdinand Bauer, it has been described as 'Oxford's finest botanical treasure'. It was a publishing phenomenon, one that would change the face of botanical research, art and gardens for ever.[7] It's a trove replete with a thousand jewel-like watercolours; together they capture dreamy landscapes and a kaleidoscope of Levantine flora.

But the *Flora Graeca* is not the whole story. These beautiful works are the fruits of an ancient Greek botanical heritage. Aristotle (384–322 BCE) and Theophrastus (*c.*371–287 BCE) were philosophers and scientists of Greek classical antiquity, described as the founding fathers of zoology and botany respectively as well as the co-founders of biology.[8] Dioscorides (*c.* CE 40–90) was a military doctor for Emperor Nero who, based on his botanical expeditions, wrote a pharmacopeia describing over 600 plants, their habitats, methods of preparation and medicinal uses.⁹ Sibthorp's seminal text was inspired by the work of these early Greek philosophers and scientists. He compared the specimens he found with the early scientists' original descriptions and was intrigued to know how people used the plants. While many of the region's plants already featured prominently in eighteenth-century medicine, little was known about the flora as a whole, the habitats of the plants or their taxonomy. Sibthorp set out to remedy this on his adventures across the eastern Mediterranean.

> *I walked out with a shepherd's boy to herborise; my pastoral botanist surprised me not a little with his nomenclature; I traced the names of Dioscorides and Theophrastus . . . their virtues faithfully handed down in the oral traditions of the country . . . such a rustic repository of ancient science.*
> Sibthorp, 1795.¹⁰

Sparkling blue arcs, wisps of orange blossom, echoing cicadas and, in spring, a floral chaos. I knew it all well. I'd taught Mediterranean field botany for a decade: marching scorched students into the gorse, peering through hand lenses under the burning sun, ducking under the knocking carob tree branches and pulling out briers from our clothes and hair. You have to be in the vegetation to understand it, we'd tell them. Then, in carrying out research for Kew's field guides to Mediterranean wild flowers, I followed the footsteps of Sibthorp around mainland Greece, Crete and Cyprus. Down wild and empty tracks into canyons I went, into fishing boats and onto islands I hopped and into the Aegean I dived, chased by the ghosts of Greek deities, mythological creatures, and by dragons.

MACEDONIA

A windy blast of aircraft on the tarmac, kerosene and a hint of something else hits my senses: spring in the Aegean. As I make my way to the arrivals lounge of Thessaloniki airport, I feel that little thrill I always get in anticipation of a botanical adventure. Anticipation and a small pang of anxiety at the prospect of driving on unknown roads in a foreign land. I'm right to be anxious because the moment I set off in my little hire car along the road east, Zeus, god of sky and thunder, unleashes his wrath. I can scarcely see through the torrents cascading down my windscreen and pooling on the road. Blue road signs listing indecipherable place names flash past me in a wet blur. And then a bolt of lightning slices the sky to give my anxieties another good dose of oxygen.

Eventually I reach a coastal town labelled Ριβιέρα ('Riviera'), park the car and collect myself. The rain stops as abruptly as it arrived, and I can see now that the town is eerily quiet. In fact, it looks abandoned. I rub the steamy car window with a squeak and peer at my lodgings – a white, boxy building with an air of the Eastern Bloc. A big sign saying 'HOTEL' and hanging at a jaunty angle looks promising, but the shutters across every window do not. I lock the car (parked at an angle mirroring the hotel sign) and explore. I locate the doorbell by my phone's torchlight and wait. *And wait.* Minutes later, a light clicks on and the door is opened by a stout lady in slippers, breathing deeply. She looks at me blankly. 'Err, guest?' I venture. She considers this, then waddles off and potters about backstage for a while, then returns with a bunch of keys (relief). We set off to my room, which is cell-like with twin beds arranged, strangely, with their headboards against the window, looking out onto a blank wall. Famished, I enquire whether there might be somewhere I can buy food. After a comical exchange of hand gestures she brings me two very deep-fried chicken legs and two very welcome beers. I settle down for the night with my copy of Oleg Polunin's

Flowers of Greece and the Balkans and a skulking cockroach for company.

Things aren't so bad. I wake up to warm Mediterranean sunshine on the back of my neck (naturally, because the bed faces the wall) and rise with a skip in my step. I'm brimming with the delicious anticipation of a two-day adventure to locate a plant I've wanted to find for years. After a brief stroll about the pleasantly ramshackle town, I set off impatiently for the foothills of the Rhodope Mountains.

The road skirts rocky mounds of maquis to my left and a pool of silent blue Aegean to my right. After an hour, I take an empty road that snakes its way into the tawny hillside of Macedonia – mile upon unspoiled mile of it. I pass blood-red chalky embankments and little rectangular fields overlooked by derelict farmhouses. I do not see a soul for hours. As the road rises the fields disappear, and a mantle of thick scrubby green vegetation envelops the hills. By the time I pass Κεχρόκαμπος (Kechrokampos), the lane narrows and rocky bald patches appear. Little purple anemones – Macedonia's harbingers of spring – brighten the wooded roadside. By lunchtime, I reach my destination: the forested hills surrounding the Fortress of Κομνηνά (Komnina), an ancient castle overlooking the valley of the Nestos river.

> *The season of spring was now commencing, and every*
> *patch of grass was covered with anemones of the most*
> *vivid hues, scarlet, white and blue; these were intermixed*
> *with the crocus, asphodel, hyacinth and purple orchis.*
> Sibthorp, 1794.[11]

I slip into the forest with my map. The deciduous trees are still bare, but the first spring flowers are emerging. I interrupt a herd of tinkling, milky-brown goats pouring into the path. We stop to stare at one another. Delicate flowers of yellow primula and little pink *Corydalis solida* poke out of gullies and little yellow stars-of-Bethlehem (*Gagea*) shine brightly out of the bare earth. It feels as though I could walk for weeks seeing no one in this remote tract of forest. I pause on a rocky ledge sprayed acid-yellow with *Euphorbia myrsinites* and soak up my first few moments of warm spring sunshine. It's a place of deep calm. Supressing the urge to nap I press on, keen to find the plant I'm seeking before dusk.

It isn't hard to miss. I turn a corner and suddenly, there, on a steep bank of hazel copse, are hundreds of them: giant parasitic toothworts (*Lathraea rhodopea*). All around me, they are pushing their way out of the leaf litter like oversized yellow asparaguses. They resemble their North European cousin, the common toothwort (*Lathraea squamaria*), only several times bigger. Each fleshy white stem is as thick as a finger, jammed with pink and yellow flowers in a spike extending well above my knee. They are not pretty. But their leafless forms give them an unearthly, ethereal aura; a static beauty. As I leave the forest and look back at them, they seem to illuminate the forest floor like church candles.

MOUNT OLYMPUS

A gargantuan tower of 52 peaks slashed by gorges, 2,917 m (9,570 ft) high with a circumference of 150 km (93 miles): Olympus is one of Europe's largest mountains by topographic prominence. And it's awe-inspiring. Its isolation, proximity to the sea and its position at the crossroads of central European and Mediterranean floras have all contributed to the evolution of a unique assemblage of plants. Mediterranean, Central European and Balkan species all exist in unusual proximity; meanwhile alpine species descend far below their typical haunts around the bleak summit peaks, into the sheltered, humid limestone valleys and canyons. The mountain is a living library. I approach the botanical monument from the northern coastal plain and the rugged grey mass rises in front of

me like a wall. Home of Zeus – god of storms – I see apposite black clouds collecting ominously about his roof.

I stop at a site at the foot of the mountain that a local naturalist tells me is good for bee orchids (*Ophrys*). Bee orchids' 3D furry little flowers delightfully resemble insects, something which perplexed Charles Darwin. He was so intrigued by pollination in orchids that in 1862 he dedicated a whole book to them, *On the Various Contrivances by which Orchids are Fertilised by Insects*.[12] However, the particular bee orchid he was observing (*O. apifera*) in England was self-pollinating – an evolutionary consequence of a divorce with its pollinator in the Mediterranean, the long-horned bee. It was not until 1927 that the Australian naturalist Edith Coleman showed that some orchids mimic the scent and appearance of female insects to trick males into attempting to mate with the flowers.[13] Amorous male bees, thrusting against the flowers, inadvertently pick up and deposit pollen, all to the benefit of the plant and receiving no reward in the process (known as *pseudocopulation* in the science biz). Myriad different forms of these peculiar little orchids grow around the Mediterranean and the Aegean is their capital.

I park on the gravelly peneplain. Within minutes I stumble across my first soggy clump of orchids. Soon, I find hundreds of them poking out of the dewy turf. Most are the common early spider orchid (*Ophrys sphegodes*), both in its typical form and the subspecies *mammosa* with horny protuberances on its flower lip. Some belong to the distinct subspecies *spruneri*, which has a prominently three-lobed lip; others still seem intermediate among the three. I know from experience that where different forms of bee orchid grow together, unfathomable specimens sometimes turn up. Perhaps this is why botanists and orchid fanatics love them. Musing over my convolution of orchids, a flash of maroon catches

Lathraea rhodopea in the Rhodope
Mountains, Macedonia.

my eye through the gloom. I dart over and, to my delight, find a splendid pair of *Iris reichenbachii* flowers in full bloom, almost a month earlier than usual. Their perfectly-formed, blackcurrant flowers are box fresh and beaded with raindrops: they are like a pageant of purple, against the seat of the gods.

I arrive at Λιτόχωρο (Litóchoro), a quaint town nestled at the foot of the mountain which is the staging area for its climbers. My guesthouse vanishes in a sea of red roof terraces and telegraph wires. Behind it, ghosts of Zeus's storm hang thickly around the complicated summit. I saunter around the town, which feels sleepy because summiting season is still weeks away. Old men are drinking ouzo in a sunny town square in the wake of the storm. I find a restaurant along the main road where I enjoy a Greek salad, washed down with a few ouzos of my own (*when in Rome . . .*). Chatting to the bored-looking but friendly barmen, they tell me the mountain is ablaze with flowers in the spring, which sounds promising. Back at the guesthouse I flick through my photos and sketches from the day and feel a sense of satisfaction wash over me – a heady cocktail of uncomplicated love for orchids and irises, mountain views and half a dozen ouzos.

I ENTER THE Enipeas Canyon on the mountain's eastern flanks by a narrow path that saddles two forest-crested grey crags and runs parallel to a crystal clear stream. It's just after sunrise and besides

A bee orchid (*Ophrys sphegodes*, *spruneri* form).

the ghosts of lesser gods and Hellenic spirits, I have the mountain to myself. Sunshine sparkles, and everything looks alive and shiny with morning dew. Khaki patches of forest clinging to the rock are dominated by a mantle of papery kermes oak (*Quercus coccifera*) and green olive (*Phillyrea latifolia*). The path takes me over rain-blackened wooden bridges, past waterfalls spilling into jade-coloured pools and out onto wrinkled spurs of rock. I wander out onto an anvil of cliff where I have an eagle's eye view over vast, fissured walls dropping steeply into the ravine. Quietude. The kind that can be found only on a mountain. At my fingertips, I notice crevices in the rock packed with violet-coloured cushions of *Aubretia thessala,* growing just as they would from an old wall in an English country garden. The path dips and rises through a dense forest where I spot a fire salamander crawling slowly at my feet. I could hardly miss it. It's jet black with yellow, wasp-like markings. I've read that these animals are long-lived; I wonder for how long this one has been in this mountain wilderness, a place that resonates deep time.

By lunchtime, I reach the crags of Priònia. Butterworts are reported to grow on the mossy boulders by the stream, but I can't find them, so press on. There are plenty of other plants to see. The ledges of the steep, folded walls at this altitude are home to a unique assortment of rarities including the endemic, yellow-flowered broom (*Genista sakellariadis*), which is found only on Olympus, and little domes of saxifrage (*Saxifraga scardica*) that usually cling to the naked rock of its highest slopes.

Eventually I reach the snowline. A yellow, wolf-like dog appears from nowhere and accompanies me along the track. He trots ten paces ahead of me then sits looking at me, panting. He looks expectant. Perhaps, like me, he has a sixth sense, can divine things. Because a few metres above where he is sitting, on a great

Wild irises (*Iris reichenbachii*) growing
at the foot of Mount Olympus, Greece.

Ramonda heldreichii, 1,100 m (3,609 ft)
up Mount Olympus, Greece.

mossy promontory, is the very plant I've climbed the eastern shoulder of Mount Olympus to see: *Ramonda heldreichii*. I scan the rock, tracing a route to the plant with my eye. I haul myself up onto the craggy mass, feeling my weight in my arms and tapping the rock with my feet to find the footholds. I watch the dog trot away down the track, disinterested. After a short scramble, I reach a broad-topped peak and a vista of rocky walls and forest unfolds around me. Here I find about a dozen rosettes of the plant growing out of a vertical chink in the rock. Their soft grey leaves feel like crushed velvet. 'Beautiful,' I whisper. I'm holding conversation with the past: *Ramonda heldreichii* is a relict from the Tertiary period that shifting conditions confined to the higher slopes of this great mountain long ago. Long predating the arrival of people, I should think it will outlast us too, up in this remote, changeless place. I rest for a while as my pounding heart slackens and take it all in – the views, the plant and the timelessness. Tens of millions of years behind it, a thousand metres of rock beneath it, here is a plant that grows to the heartbeat of Mount Olympus.

We descended a rocky precipice, crossed a stream,
and preceded along a narrow glen, walled in by high
mountains covered with forest trees; and from the sides
of which issued forth cascades and purling rills.
Sibthorp, 1795.[14]

CRETE

'Like new, yeah?' beams the hire car assistant, handing me the key to the vehicle outside the airport car park. He's not wrong – the gleaming white Polo is pristine. Which is unfortunate, it turns out.

It has rained for days and everything glistens under the soft evening sun. The air smells of wet terracotta. I read the landscape as I drive to Chania. Its grassy roadsides are a vision of floral excess

– splashes of red, blue and white, buttery annuals sprawling out onto the tarmac, with clouds and great stands of lemon-yellow giant fennel (*Ferula communis*) the size of people. Spring has arrived on Crete. I feel sweet anticipation.

The promised parking space in Chania doesn't exist and I become swallowed up by a complicated one-way system designed for vespas. Without an obvious alternative, I set off up a narrow, cobbled street. Old boys sitting with their backs to the yellow and red-painted walls drag on their cigarettes, staring intently as I trundle uphill in second gear. Sword-like leaves of potted yuccas rasp and thud the car windows aggressively. The narrow street ends at an impossible T-junction into which I turn right and (small mercies) achieve the most advanced fifteen-point turn of my life, ready to return down the impossible street. Only now that I'm facing the opposite direction, I can't seem to reorient the vehicle. The last time I'd this much trouble with a car was when I rescued a two-metre-long cactus destined for the compost heap that stabbed my knuckles every time I shifted gear. I rev the car noisily, attracting the attention of a gang of teenagers. Excruciating beads of sweat trickle down my back. If I managed to get it up here, I *must* be able to get it out again. I allow myself a small smile, convinced I'll *just* clear it; and then I hear the ear-splitting shriek of metal grinding plaster. My left wing mirror snaps. Not like new any more . . .

OF THE 1,400 ISLANDS sprinkled across the Aegean Sea, Crete is the largest. Largest and richest, botanically speaking, its bewilderingly diverse flora has always piqued the curiosity of botanists. The island is a vestige of an ancient mountainous arc that once spanned mainland Greece and Turkey, and it still possesses remnants of their respective floras. A legacy of continental connections to Europe, Asia and Africa, followed by five million years of isolation, have blessed the island with over 2,000 plant species. Two of its genera are unique to Crete and have no close living relatives. To the field botanist, this unique assemblage of plants bursting out of every crevice of Crete's rocky slopes, gorges and ravines during March and April is very heaven. And I'm in the thick of it.

I head to the Gramvousa Peninsula, the island's north-western promontory, to examine the coastal *garrigue* flora. The road west of Chania is edged with yellow clouds of acacia, pink-spotted oleander bushes and rusty cliffs bespangled with wildflowers shimmering in the breeze. Shards of glittery blue sea flicker between the ranks of cypress trees and the soapy smell of orange blossom wafts through the car. I can imagine the pulsing buzz of cicadas in midsummer. Taking a right turn, I crackle down a white gravelly track, then another, until that peters out by a notch of shingle in the sickle-shaped peninsula. I abandon the car and set off on foot down a lonely yellow footpath that mirrors the complicated contours of the coastline. After about 40 minutes, a sea fret pours over the land, softens the sun and casts an argent stillness over the sea. Above me, great scoops of apricot rock bristle with domed spiny bushes. And in every fold of rock, little flashes of colour give away botanical treasure.

You can't swing a cat on Crete without hitting an orchid. The island is a cornucopia of them. Soon I spot my first: a pink butterfly orchid (*Anacamptis papilionacea*). This one is quite common on stony slopes across the Mediterranean. The specimen in front of me is small and has four flowers, with only one in its prime. I peer at it through my hand lens to get an insect's eye view. It has a pink, frilly, heart-shaped lip decorated with felt-tip-thick dashes; at its centre – the business part of the flower

– its 'bits' are configured in an uncomplicated pink smile and its three fuchsia-coloured sepals point forward in an embrace. All things considered, it's a happy-looking thing. By the afternoon I approach the end of the peninsula, which is covered in starry white asphodels (*Asphodelus ramosus*). I've read that in Greek legend asphodels embodied the underworld, perhaps because of their deathly grey and white pallor. The hillside in front of me, dropping steeply to the sea and covered thickly in them, is a spellbinding prospect.

Back in Chania, a wind has whipped up. White awning flaps around noisily and waiters are losing the fight with paper table coverings against a restless grey sea. The shutters rattle. Another storm is brewing.

TODAY I'M VENTURING inland to examine Crete's unique gorge flora. I wind my way up into the mountains, past white-lime houses flung about the hills. A woman taking down washing, pausing with a hand on one hip; a pair of old women dressed in black, pushing a reluctant donkey up the lane while a man looks on, bored. People slowly going about their business under the bruised sky.

The aroids of Crete. Clockwise from top left: *Arum ideaum, A. cyrenaicum, A. creticum, A. purpureospathum, Dracunculus vulgaris, A. concinnatum.*

I leave the car at the foot of the folded grey cliff and slip down a mule track marked *Imbros*. I'm out of sorts today. I kick at stones on the road. It should be warm and sunny in the Mediterranean in April. And things should be in flower. But everything is late this year because of the cold, wet spring. Perhaps my bad mood results from two weeks' solitude. Then again, perhaps it's because I'm heading deeper and deeper into a remote gorge in the filthiest weather imaginable. Vicious gusts fling sheets of rain from behind me that fall and echo from the earth. Empty-tasting rain trickles down my face and clings heavily to my clothes. I could be in Scotland in February. (How long does pneumonia take to develop?). I press on disconsolately with my head down, pausing to examine a cluster of bee orchids growing by the rain-blackened path. They might cheer me up. Their furry flowers are sodden. They look like angry wet cats. Some are difficult to put names to at the best of times, but in this condition, identification is just about impossible. Being of a scientific turn of mind, the nagging part of my brain that sees security in detail – likes everything orderly, defined and classified with the loose ends tied up – feels unsatisfied. Their markings sneer at me tauntingly. I scowl back at them as I pull up the collar of my cagoule and put my hood up. Vexatious orchids. Spiteful rain.

The wind blew fresh from the island . . . but though I found some thousand plants I could not meet with a single specimen in flower.
Sibthorp, 1787.[15]

A sawfly orchid (*Ophrys tenthredinifera*), Crete.

By lunchtime the storm has chased away the worst of the clouds and my mood, and everything glistens with life again. The rain has unleashed fresh smells from the rock, like old books, and insects begin to stir in the growing heat. For the good of my soul, I spend an hour bathing in the warm sunshine among golden spires of yellow asphodel. It's not a bad place in the world to be. Like mountains, gorges are among the few places untouched by human progress, where plants hang on through the eons. Yellow-flowered tree flax (*Linum arboretum*) sprays out of the rocky crags above my head. The gentle bouldery track winds through narrow, shaded gaps and out into rocky screes, waxing and waning to the coast. Near the mouth of the gorge, on the way to Hora Sfakion, I see dragon arums (*Dracunculus vulgaris*) still tight in bud. Their long spadices are held aloft on solid-looking purple-spotted stems. Each flowering structure is wrapped tightly by its spathe (a leafy, sheathing bract-like structure) and tilts skyward. I search for one in full bloom but, frustratingly, they are all still a good two to three weeks from opening.

On the return leg, I pass a naturalists' tour group scattered across the path, peering this way and that. One of the party is staring up at something through a pair of binoculars. Behind her, I notice an octogenarian, apparently photographing her bottom; on closer inspection I notice a reed to which an exotic-looking insect is clinging, which is perhaps in his line of sight. I chat to the party for a while and we compare notes about the plants we have found in flower, or struggled to, in this uncharacteristically cold spring. On discovering that I'm a botanist, one lady asks what my favourite plant is. People always ask this. You have to seem cooperative don't you, so I make something up. The truth is, I don't have one, two or even a dozen favourite plants. Today it would have been a dragon arum had they been in flower. Wearily, I make my way back to the car.

In the evening, Chania's maze of streets leads me down to the harbour where tourists with blonde hair and white linen clothes are milling about, looking at menus in a cloud of perfume. The sea is calm again, its surface sprinkled with winking shards of orange light. Dreamy. I find in a bar where I can sit and think. I drink beer and turn over the day's minor lunges and reverses in my mind, pushing out the occasional tweet about the plants I've seen, validating my existence to the world. Tomorrow I plan to head to a sun-baked cliff on the south coast. It will be my last shot at seeing a dragon arum in bloom this trip. The lady in the hotel told me that the x on my map is inaccessible on foot and I'd be hard pushed to persuade a fisherman to take me there. But it'll work out.

The storm increased, and prevented us from examining further. We took shelter in a natural grotto immediately under the summit of the mountain; some alpine plants, now in flower, hung pendant in festoons from the walls of it . . . the storm now increased to a hurricane; the wind raged; the hail drove furiously along the mountain; we with difficulty kept our feet on the declivity of the rock; loose stones covered the road, and made our progress dangerous and uncertain. We arrived . . . deluged with rain.
Sibthorp, 1787.[16]

AT BREAKFAST I drink my coffee down to the sludge and scrutinise my map. I've been tipped off by a local bird watcher on Facebook that dragon arums are in bloom on the very south of the island; he thinks he's seen them on a remote cliff through his binoculars. Only there's no road or path to their location. Perhaps the fishermen down at the village know of a route to the area that is accessible on foot? Fishermen know everything about the coast, don't they? 'Pah! You won't get an answer from them,' disputes the breakfast lady over her shoulder, her arms laden with plates. 'All they're interested in down there is their fishing boats.'

'It's hard *work* on a fishing boat,' says the old, brown-skinned fisherman down in the village, busying himself with his nets on a sun-rotted peeling blue boat, avoiding eye contact. 'People don't understand that. Storms. No fish. It isn't easy.' I nod empathetically. He's not in the mood for maps and dragons, I can see that. So I return to the car. 'Your route contains unpaved roads. Do you wish to continue?' sneers the automated voice in my GPS. No one seems particularly optimistic about my adventure. I ignore them all and drive up to the top of the cliffs to get my bearings. I'll find a way down – I always do.

A clear blue day. I scan the horizon from the top of the cliff, its rocky precipice as warm as blood under my hands. Beneath it swoops a view over a steep, honey-coloured slope scattered with hunched shrubs, which levels out towards the sea. Bits of land break off into island fragments that drift out over the horizon. A warm gust rushes up the hill, rising and falling, stirring the bushes. The sea looks calm, and a blue too perfect for any camera to capture. Early April on the heat-struck south coast of Crete feels like a fine day in July. I drink it all in, file away the memory to trawl up on a bleak winter day.

Take your time – that's the trick. You mustn't rush down a cliff. Cheek to the rock, legs dangling, I gently kick the cliff to feel the footholds – talk to it through my feet. There are plenty of them for the surface is so uneven. The trouble is, they keep vanishing. For no sooner have I pushed my boot into a crevice, the rock crumbles in a stream of dust. *Stay calm.* I've all the time in the world to reach the bottom. Even so, I feel my fingers fizz with adrenaline and clammy droplets of sweat seep through my palms – inevitable consequences of so much air beneath your feet. It's exhilarating though. Trust me, it is. What's wonderful about cliffs is all the plants that tough out a living on them, soldiering on through the seasons – in fact, just a couple of metres to my left I can see . . . *Focus.* The sun beats down on the back of my neck and outlines every detail in the rock in high definition. Down, down, down I go. I kick, twist, dangle and hop the last two metres, and at last I tumble to the bottom. All things considered, I've had worse journeys down a cliff. I brush off the dust and look around me.

You smell a dragon before you see one. And to inhale a whole herd at full power, at the foot of a cliff on a hot day in Crete – that's something you don't forget. It's like crawling through a sewer. But a thunder of dragons is an arresting sight. Each flowering spike stands to a commanding one metre or more. You find these extraordinary things grown in gardens as curiosities, but seeing them wild, projecting out of the rock against the glittering sea, is a spectacle. Little wonder they were associated with snakes, magic and aphrodisia in the Middle Ages. I make my way around the herd. A confident-looking specimen has a frilly-edged, velvety spathe flung back like a dog's tongue, from which a wrinkled, shiny spadix points to the sky. It's the colour of dark chocolate. I hear the stop start drone of flies. Entranced by the noxious aroma, they are duped into believing they have found a rotting carcass of a goat to lay their eggs on – or, in this case, a whole massacre of them. Upon investigation, buzzing from one flower to the next, the flies unwittingly bring about their cross pollination. I watch one crawl out of a flower at the peak of its powers, clinging with pollen; see the event unfolding in front of me. Evolution in action. I savour the dragons in the shimmering heat for as long as can stand their fierce breath, then clamber down to the sea to cool off.

I drip dry, darkening the sun-blistered terracotta rock around me like a rash, my scratches stinging with salt. Rough, white streaks of salt have formed bands across my darkening skin. I can hear the gentle sigh and hiss of the slack tide beneath the rockfall. Out on the horizon stretches a snout of headland sprinkled with cuboid white houses. I gather my things and set off down the little pitted track to the village that dips and rises, winding steeply about the cliff, not leaving the sea. Little tongue orchids and corn chamomiles flower against the white sparkle.

After an hour of sea, light and nothing else, I detect signs of habitation: the drone of a motorbike; the grind-*bash* of a circular saw; undersized cats seeping into the wall like quicksilver. An octopus pegged up to dry alongside the family washing. And a gnarled old woman, dressed in black, climbing painfully up a stepped street that smells strongly of laundry powder, while a radio blares from behind the shutters. But, despite the low hum of the place, it's devoid of people. It has the lassitude of sleepy villages across the Mediterranean. I find a weather-beaten restaurant next to the flashing water. It smells of boats and petrol. A line of dark, sombre-looking builders sit along the wall smoking, while a yellow dog sleeps fitfully in the whitewashed shade at their feet. The waiter speaks in Greek. It pleases me that I've passed for a native, makes me feel at one with the place. But the illusion is quickly undone by my arrant incomprehension and he switches to German; in English, I order a Greek salad that I wash down with two beers. And a third. It goes to my head in the dizzy heat. Lazy afternoon. Rather unsteadily, I retreat to the hills and nap in the web of shade of a lonely carob tree, dreaming of cliffs, sea and dragons, in the sun.

Gradually the magic of the island settled over us as gently
and clingingly as pollen.
Gerald Durrell, 1956.[17]

ELEFTHERIOS DARIOTIS, a native plantsman, has drafted a battalion of shepherds from across the island to find wild peonies in flower. We look out for each other, botanists. But they are late. I hunt for them all over the Omalos Plain, hauling myself through wire fences, peering around abandoned buildings and weaving through wintery rows of bare trees backed by snow-streaked mountains. But it's no use: if a shepherd can't find one, I know there's little point in looking. So I change my plans and head for the Kedros Mountains in the centre of the island where the flowers are more advanced. I squint at my oversized map spread out on the dirt and trace the road east. Before I set off, I take the opportunity to plaster the side of the car with wet mud to hide the evidence of the altercation with the wall from earlier in the week.

The road inland is wrapped in butterscotch cliffs and stony slopes on which bent figures are harvesting something with large baskets beside them; possibly snails, for it rained last night. People are attuned to the land here, its riches and its rhythm. The land drops away from the road. A pair of honey-brown eagles glide round an unseen staircase, tilting, with their eyes fixed to the ground. Sunlight flashes about their wings. I stop the car to examine a clump of Cretan arum lilies (*Arum creticum*) with glossy, spear-shaped foliage, poking out of the sherry-stained rock. Unlike their distant relative, the dragon arum, these lemon cheesecake-coloured blooms smell sweet, like lilies. Just a short walk from them I bump into another member of the family: *Arum concinnatum*, growing in the green shade of the undergrowth. Like its cousins on the cliffs, it plays a rotten trick on its pollinators: this species produces a scent like animal dung to attract midges and, to perfect its mimicry, its yellow spadix is even warm to the touch, just like fresh manure. The little midges zoom into a floral chamber where they become trapped by a barricade of spines and are showered with pollen. The following day, these spines wither to release the insects, some of which may visit another arum and bring about cross pollination. I sketch the wand-like yellow spadix flopping out of its spathe.

I wind my way up to the Spili Bumps. This botanical hotspot of lush plains and limestone hills is famous for its diversity of bulbs and orchids. I park next to a yoga retreat where around 20 participants are bending and swaying in synchrony on a hill framed by steely-grey mountains. A stone's throw from the car, I spot clumps of wild irises that have me hopping about with excitement: widow irises (*Iris tuberosa*) with vellum petals in watercolour blues, blacks and yellows, and clumps of Algerian iris (*Iris unguicularis* subsp. *cretensis*) the colour of night skies with yolk-yellow centres. I try to capture them on camera; this proves to be difficult because of the brilliant sun pointing in one direction and the yoga participants' bottoms in the other. I take a track splicing lime-green fields and rough grey mounds, out onto the Gious Kampos plateau. Here, in the shadow of Mount Kedros, I enter a glade of blood-red tulips (*Tulipa doerfleri*) bobbing in the breeze. This tulip is found nowhere else on earth; here, and here only, in this little corner of Crete, grows a whole ocean of them limned by silver strips of mountain and sky.

A Cretan bee orchid (*Ophrys kotschyi* subsp. *cretica*), Crete.

It's such a special place. The limestone bumps are like puddles of abundance in a shimmering sea of grasses: orchids in pink pyramids (*Anacamptis pyramidalis*), raspberry ripple clouds (*Orchis italica*), splashes of lemon meringue (*Orchis pauciflora*) and a swarm of sawfly orchids (*Ophrys tenthredinifera*) all jostle for position among drifts of white Mediterranean hartwort (*Tordylium apulum*). These bumps must be as diverse as the equivalent area of a rainforest. As if in a dream, I float through a constellation of wildflowers, past crumbled ruins and clouds of almond blossom up onto a rocky scree, where I look out over the orchid realm. Presiding over it all, out of the rock, stands its sovereign: a solitary Cretan bee orchid (*Ophrys kotschyi* subsp. *cretica*). Its velveteen black lip is slashed dramatically with silver and white bars. Its high contrast geometric markings look meaningful, like a cryptic code, a symbol of the Phoenician alphabet, or something like that. It's a singularly mesmerising work of nature. I stare until my eyes lose focus and brain can't keep the colours constant. I've been hypnotised.

CYPRUS

The lazy drone of a bee looping in and out of pink flowers shimmering in the breeze, grey hills with a fresh green cast, and an ocean of orchids. April on the island of Cyprus. A tailed comet, set adrift from the Levantine mainland; a botanical wonderland floating off into the Aegean. This is where you'll find me, and there's nowhere else in the world I'd rather be today.

Down below the island's north-westerly horn, I sit cogitating over some early spider-orchids (*Ophrys sphegodes*), growing in the long grass of a roadside pasture. Their furry little 3D flowers have a spellbinding exactitude. A wizened old man wanders up the lane and shouts amiably at me, breaking the spell. He is curious to know what I'm looking at, so I point out the various orchids I've found in this little green parcel of vegetation. He speaks in Greek until it becomes clear that I can't understand him.

'Turkey?' he asks.

'I'm from the UK.'

He peers at me as one might look at an exotic animal in a zoo, laughs long and hard, looks up to the sky, repeats 'UK' and clenches his stomach with a brown, knotted hand. Then he turns serious again. 'But you have the Turkey face.'

'Well . . .'

'You like Cyprus.' He says in a deep, heavily accented voice with drawn out vowels, nodding, agreeing with himself.

I say I do, yes, and by way of explanation point out more orchids I've found.

'Meh. We have better ones than that! Big pink one. Yellow one. These? Small, brown thing,' he says dismissively, waving them all away.

We have a charmingly illogical conversation about communication through power lines, the seasons in Cyprus, the rain in Great Britain and the economy – just generally, punctuated with volleys of laughter. He makes a broad sweeping hand gesture and concludes, 'but that's just the way things are'. Slowly, he climbs back up the lane, leaving me to ruminate over my little brown orchids.

I have lunch in the netted shade of a fig tree at a roadside gaff made for card players, drinkers and old boys with an afternoon to waste; then I head north to Akamas, the island's desolate, western promontory.

IN THE SHADE of the eaves of a crooked old fig tree, water trickles down a wall of rock into a misty pool of Prussian blue. They have an enchanting beauty, the Baths of Aphrodite. The goddess of beauty, she met her lover Adonis at this pool when he stopped for a drink while hunting, according to legend. And, if the leaflet is to be believed, 'the spot fills the visitor's soul with soothing calmness and tranquillity'. But not today. Today it's occupied by tourists gleaming in white jeans and aviator sunglasses, re-enacting Aphrodite's poses and taking selfies by the pool. One tourist nearly topples in with a little scream (I shouldn't laugh). I poke about the vegetation around the pool for a while and then head out to explore the wild and rugged peninsula.

Dusty iron hills, scattered with awkward pines casting spidery shadows, drop steeply to the sea. I trudge softly through a mantle of spent needles that fill the air with a smell as strong and sharp as disinfectant. Here and there, wild cyclamens (*Cyclamen persicum*) form lakes of white shuttlecocks from which jagged rocks jut out like islands. A few metres beneath the pitted track, brilliant blue water wobbles with a thousand reflections that flash white along the rock. Then the path leaves the sea and winds steadily up into barren wilderness.

Phrygana is the stunted, sun-drenched form of vegetation so characteristic of the eastern Mediterranean. The oils of countless aromatic herbs all volatilise together under the searing sun to create a heady cocktail of fennel, thyme, rosemary and sage. It's a gift to the senses. In spring, it's a gift to the botanist too. For a short time after the winter rain, this rich, terra rossa earth whispers with a thousand plants. There are so many I scarcely know how to begin examining them all. Confirming the identity of every species with Meikle's weighty *Flora of Cyprus* slows me down. After progressing just five paces in 20 minutes, I prioritise my time and focus on the plants unique to the island – the most special.

Speaking of special: enter Cyprus tulip (*Tulipa cypria*). This exceptionally rare bulb is reputedly difficult to strike, but just a few metres from the rust-coloured crooked track, in the scrubby undergrowth, I find one. Its sleek, pleated petals are precisely the colour of ripe cherries. Mindful not to tread on any emerging buds, I crawl on all fours to look at it. Gosh, it's a beauty. The next plant of interest I find along the trail is a tongue orchid (*Serapias politisii*), of the rare *aphroditae* form, growing out of the red clay between the apricot slabs. Now here is a peculiar plant. The flower of a tongue orchid is folded to form a little tunnel that mimics the protective crevices sought by nesting bees. As they wriggle in and out of these floral tubes, they act as couriers for the plants' pollen. In other places in the Mediterranean, I've located the little insects lurking inside the flowers. Here the lodgings are vacant, but for the single gossamer thread of a phantom spider.

I look out over the long, curved spine of the peninsula, its rocky vertebrae repeating themselves to the sea. Khaki slopes drop gently from either side to an intricate white coastline of sand-strewn little coves and bluffs, trimming an empty blue sea. The scrubby thicket that overlooks it is splashed with chrome yellow Persian buttercups (*Ranunculus asiaticus*), a plant frequently encountered in the eastern Mediterranean, each locality inclined to its own particular colour form. Here creams and yellows dominate. Far less conspicuous than these, at my feet, I spot spikes of *Bellevalia nivalis* no taller than my index finger, unfolding furtively from the red raw earth. Its flowers are the colour of ice. From a crevice a few metres beneath them sprouts its larger relative, *Bellevalia trifoliata*, with spikes as long as my forearm carrying unevenly placed flowers, topped with buds the colour of the sea. Together they hint at the rich botanical treasure that hides in every pocket of this land.

Having employed the morning in drawing, and putting our plants in paper; we rode out after dinner to the monastery of Lapasis, a fine remain of old Gothic structure . . .
The sun shone with uncommon force; nor did the least breeze mitigate the fervour of its rays . . . We joined our

companions at the monastery of Lapasis, situated in a
beautiful recess, surrounded by corn-fields and vineyards,
and shaded by a tree, whose foliage is kept green by several
purling rills, that watered the environs of this romantic spot.
Sibthorp, 1794.[18]

A PLANT THAT lives inside a plant. Curious, don't you think? I've always thought so. There are just four examples in the plant kingdom, the so-called endoparasites, and I'm kneeling beside one on the rocks. I've stumbled across the peculiar clusters of *Cytinus*. It's covered in scales that look like the wax used to wrap Edam cheese. It teems with ants, possibly attracted by the flowers' faint yeasty, vinegary smell. But what's fascinating is that this plant spends its entire life living inside the tissues of another, in this case a cistus bush, from which it emerges only briefly to flower and set seed. For most of its life it exists as a network of microscopic filaments from which 'sinkers', analogous to conventional roots, syphon off water and nutrients. During the evolutionary switch from a free-living existence, plants like *Cytinus* have lost their genes for photosynthesis. More curious still, they've acquired new ones: so-called horizontal gene transfers (or HGTs in the molecular biology game) describe the rare exchange of genes in the absence of mating. And it isn't just DNA that has gone walkabouts; viruses, proteins and RNA (which acts as a messenger, carrying the instructions of the DNA) all move between parasitic plants and their hosts. They are evolutionary enigmas. Elsewhere in the world one has reached astonishing proportions – but we'll save that for another chapter.

At its narrowest point, it feels a bit like potholing, trekking through here. Probably not nearly that bad, still I've always had a better head for heights than for enclosed spaces. I squeeze my way between vast cold walls of rock the colour of wet plaster and scarred like the moon, then squint up at the blazing sliver of sky a few metres above. A stream running through the gorge has spilt out into milky-blue puddles around the path. Further in, the clinical white rock has eroded to its ribs. Ferns spew out into the wetness and trees have clamped themselves to the foot of the cliff with roots like talons, their trunks weaving skyward. The place has a quiet, secluded beauty.

As I head deeper in, the tricking stream supplants the path and my feet become drenched, cold and heavy. This is all promising, because the plant I'm looking for is called the water arum (*Arum hygrophilum*). After boring in a few kilometres, I slosh my way over to one growing among the smooth taupe boulders of a shady river bank. Its spathes have an intangible whitish-green cast, suffused at the edges with dark chocolate. Each one sheathes a matching brown, slender spadix. They have an artless beauty, something like an understated, less try-hard variant of a florist's white arum lily. It's perhaps the most beautiful arum I've seen yet.

'DANGER. STRICTLY FORBIDDEN. KEEP OUT.' The sign catches my eye from the car. They often do. The best plants tend to be found in places where people don't want you poking about. Still, this one also says '*British Forces*' and looks particularly officious, so it seems sensible to stay out. I drop down to the

Akrotiri Peninsula on the south coast to hunt down the rare Cyprus bee orchid (*Ophrys kotschyi*). Much of the land is RAF-occupied and off limits, but there are snatches I can get to. The peninsula is a sandy appendix to the mainland, half sinking into the sea and sagging with great lakes and marshes. It's an intriguing landscape, splashed olive green with bushes and netted with white sandy tracks. It is also bedevilled by mosquitoes. (Maybe this is what the sign was referring to?) They assail me from the moment I leave the car, arriving more quickly than I can possibly smack them off. I stand there swotting my legs, arms and head, jumping up and down like a man possessed, until the palms of my hands are bespattered with pressed insects and blood. Watching the drama unfold from a few metres away is a grey-whiskered shepherd and his herd. They all stare at me. The shepherd and I nod and say *yassou* to one another and then I delve into the undergrowth, curious to know how he fends off the insects. My approach is to move quickly. I zoom around the stony islands of thicket, giving no more than 10 seconds' attention to any one plant, lest the insects land on me. After 20 minutes, I find what I'm looking for: a solitary Cyprus bee orchid. Curiously, it looks as though it's smiling. I smile back, stroke it gently and then run back to the car. By the time I reach the local pharmacy I have so many bites that the staff behind the counter gasp in horror.

A Cyprus bee orchid
(*Ophrys kotschyi*), Cyprus.

FOR A MOMENT, I see myself as they must see me in the coach a few metres down on the road below: legs and arms splayed like a starfish, glued to the cliff. I've clambered up here to examine the rock-dwelling flora of the Troodos Mountains. The coach stopped in the road beneath me and now the people inside are standing up and staring. This happens occasionally. What caught my eye were the candyfloss clouds of *Arabis purpurea*. Their pink flowers form garlands around the jagged rock. The petals are the colour of apple blossom and each is delicately veined cerise. I turn one over between my fingers, looking at both sides. I can imagine labouring over the details of it with watercolour; can picture the exact brush I'd use, and which colours. The coach trundles away and I have the mountain to myself again. I hop down onto the warm tarmac and wander up the road, poking about the plants growing along the verge. Strawberry trees (*Arbutus andrachne*) lean out of denser pockets of vegetation, against overlapping blue mountain cut-outs that lift and fade into whipped cirrus. I laugh at a naked man orchid (*Orchis italica*) growing in their webbed shadows; his pink, whiskered flowers are the perfect silhouettes of winged sprites with smiling faces, bonnets and assertive little . . . (well, the clue is in the name!). Its more graceful cousin *Orchis troodi* grows here too, and I find a cluster on a mound behind the road. Their flowers seem too large for their wiry stems, like flamingos in flight.

I head to the car and consult the handmade map a local ecologist had kindly drawn for me, marking my next target. 'Not easy,' he'd said, shaking his head.

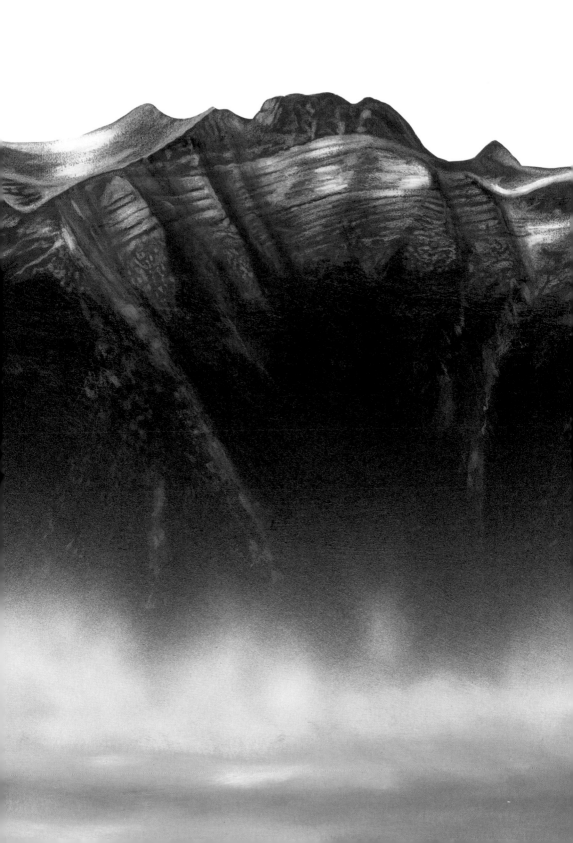

Arum rupicola in the
Paphos Forest, Cyprus.

I PEER OVER the precipice of a cliff-cum-scree that tumbles into the depths of the bottle-green Paphos Forest. Golden sunshine of a particular softness has crept over the trees and bubble-gum-pink patches of thyme (*Thymus integer*) spill out over the rocks at my feet. I have two clumps of *Arum rupicola*, about eight metres beneath me, in my line of sight. I've driven for hours to reach this remote tract of forest and I *must* see them, but the ecologist was right: it won't be easy. I scan the slope for some time and there isn't an obvious route. No path, few trees, nothing to grab hold of. Can I get down there? It seems reckless to try. I weigh it all up in my mind, wearily, at the end of a long day in the field: irresistible plant versus a twisted ankle; cuts, sprains; broken leg; potential health insurance claims . . . Yes, I weigh it up and, of course, in the end I do the right thing.

I go down.

Hell, this was the wrong decision. With a thunder clap and a hiss the earth begins to move, taking me with it. To my horror, I am abseiling down the slope. Instinctively, I clasp my shins and curl up as the rocks seethe beneath me. It all happens quickly. The avalanche slackens and deposits me a few metres south of where I need to be; nevertheless, it's a lucky escape. I gather my senses, looking at the work in store above me, then gingerly crawl back up on all fours, unleashing a further three minor landslides along the way. Lying on my front, my chin to the rock, I stare up at my prize. Their flag-like spathes are just unwrapping to reveal long, finger-thick spadices. They glint in the evening light as though lit from within, glowing red. I'm transfixed by the plants; perhaps I made the right decision, after all. I reach terra firma, scratched and bloody. By the time I leave, the dozing Paphos Forest has become torched bronze by the melting sun. I sit and watch it for a while.

DRAGON SLAYING

LAST DAY. There's one more plant I need to see here. I march through thorny bushes and grasses like a man possessed. I've got two hours before I need to check in for my flight and mustn't get distracted. That's difficult on the island of Cyprus in April, the botanical wonderland of the Aegean. I could stop and look at a dozen things I don't have time for. The slope tilts, propelling me down into a ravine. Tramping through a ditch on the way to the airport is not the most ceremonious of ends to the trip but needs must. I duck under branches, pulling twigs from my hair and thorns from my sock. I think I see one. I speed up.

There! They are just unfolding, half a dozen, with green spathes spattered all over with wine-red blotches. They have a faintly unpleasant smell about them, like a 50-watt dragon arum with a dash of manure. Oh, they are beautiful. Curious but beautiful. Their silky sheen snatches blue from the light; their spadices are freedom-seeking, splaying around like a nest of emerging vipers. They look like they might dart out at any moment. They're better than beautiful; they're *perfect*. I'm unable to stop smiling from ear to ear like a lunatic.

Now I've seen every arum there is to see in the Aegean. And that nagging part of my brain that sees security in detail, likes everything orderly, defined and classified with the loose ends tied up? It's at peace.

> *I went on shore to botanise . . . The pomegranate, glowing with its scarlet colours, ornamented the thicket; while the Vitis Labrusca, stretching over the rivulet, perfumed the air with the most fragrant odours. Oranges and lemons grew in wild luxuriance . . . I returned highly satisfied from my walk, and richly laden with curious plants.*
> Sibthorp, 1787.[19]

Arum dioscoridis, Cyprus.

4

THROUGH HOLY LAND

THE MIDDLE EAST

4

THROUGH HOLY LAND

THE MIDDLE EAST

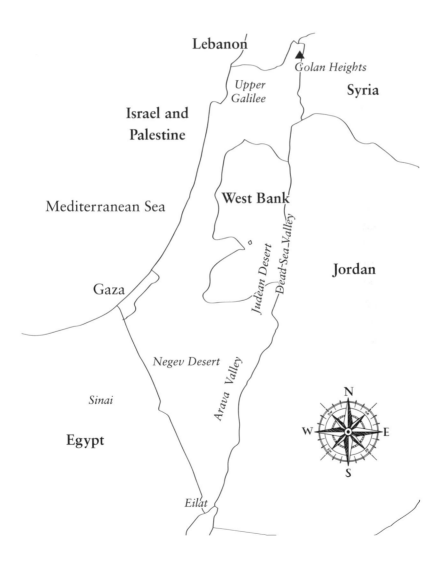

*On the crest of a hill in the Golan Heights, we find
what we came for.*

I FIRST WENT there for the irises; my lodestars to the Middle
East. I'd always been enchanted by the flowers. There's something
uniquely pleasing about their geometry, isn't there? The paired
triads of petals, one dropping beneath the other like a reflection.
The so-called Oncocyclus irises, those with the large and
exceptionally beautiful flowers familiar to gardeners the world
over, originated in the Middle East. The prospect of seeing them
there, growing wild in its tumult of mountains, plains and deserts,
was tantalising.

When we think of habitats for plants, it's tempting to neglect
deserts in favour of leafier places. The Middle East conjures up
images of dusty plains, barren mountains and rolling lifeless
dunes. So does a Google Image search, actually.

Judean *Desert*. Coastal *Plain*. *Dead* Sea Valley. The words
don't inspire much confidence in plant life, do they? They sound
like desolate, sun-barren wastes. Places left to crack, bake and
perish in the sun.

Fertile Crescent. Sounds more promising. A region of origin
for Neolithic founder crops. A cradle of civilisation where people
first cleared the land and domesticated plants. A global centre for
wild crop relatives.

A collision of continents. Sounds violent, but an important
point, nevertheless: mountains rolling into coastal plains, valleys
and deserts, together with climatic influences from all sides have
bent the Middle Eastern flora into its current shape.

A region of exceptional species richness. Let's run with this. A
land that, for just few short weeks in spring, sings with flowers.
Flowers such as blood-red tulips that bespangle the Judean Desert,
pink and white alliums that sprinkle stars onto the Coastal Plain
and roses of Jericho that breathe life into the Dead Sea Valley.

A few years ago, I received funding to accompany Israeli botanists on expeditions to collect samples of desert hyacinths (*Cistanche*), plants distantly related to the broomrapes. Imagine a broomrape on steroids. Double it. Triple it. Raise it to your knee; then stick it in the depths of the desert. That's a desert hyacinth. Our work sought to understand the biology of these extraordinary parasitic plants: their evolutionary relatedness, their taxonomy, and their distributions. The work saw me searching for wildflowers down valleys, up mountains and into deserts, chasing desert hyacinths, tulips, alliums and irises. Irises more heart-piercingly beautiful than you could ever imagine.

I'd do anything to see the perfect wild iris. Anything at all.

THE UPPER GALILEE

Mindfulness. This means paying full attention to a single moment in time and the world around you, to improve your mental well-being; or so I've read. Well, here I am, sitting on a rocky shelf in the Upper Galilee, acutely aware of the world around me. Yes, I'm mindful of the orange path beneath me, snaking alongside a stream and around green silken hills; hills netted by goat tracks, like veins of a leaf; and the wildflowers – millions of them – like puddles of sunshine. Mindfulness also enhances our connectedness with nature, apparently, and this is good for our mental well-being too. Mine spiked dramatically just now when I stumbled across the world's most beautiful iris. I'm quietly sitting with it.

The Nazareth iris (*Iris bismarckiana*) challenges our understanding of beauty in nature: what it is, what it makes you feel. It makes you wonder why we've bothered messing about with it for millennia: creating hybrids, cultivars, ornamental cut flowers and the like. Because this wild iris has an artless beauty far superior to any of that. Its three white standard petals curve upwards into an arch that has snatched blue from the sky; each gossamer-thin

A Nazareth iris (*Iris bismarckiana*) in the Upper Galilee.

vein is pencilled to perfection. Beneath these unfurl three fall petals, the colour of champagne shot lightly with gun-metal grey. And here's the extraordinary thing: they are intricately netted in dots and crosses, as if wrapped in microscopic barbed wire.

'Chris! More here, Chris!' calls Yuval from a rocky ledge below. Yuval Sapir is the leading expert on Middle Eastern irises and he's brought me to his little piece of paradise. I clamber down the grey, rocky terraces sprinkled with yellow flowers to see the next stand. Five perfect irises of ascending height are framed by yellow clouds of *Sinapis alba*, domed grey hills and blue sky. Five repeats of three-fold symmetry; 15 times more beautiful than anything I've ever seen.

Down by the path I find a patch of fan-lipped orchids (*Anacamptis collina*). The flowers have stern little faces, and arms held high. Above them towers a clump of *Phlomoides laciniata*. A confident-looking herb with fat spikes of cotton wool from which mint-like flowers are peeping out. Decoctions of the plant have been used to treat various ills across the Middle East; perhaps what I'm looking at is a relic of historic cultivation in the Upper Galilee. Further up the track, I crouch to examine the synthetic pink flowers of *Linum pubescens;* they're like jewels in the earth. Above these float empyrean white swathes of dominica sage (*Salvia dominica*), flowering enthusiastically. It's a place so bewilderingly rich in plants it has my brain buzzing.

Fullness of mind. A reversal of the art of paying attention to a single moment? It sounds as though your mind is full to the point of bursting: overwhelmed. Perhaps, on reflection, I'm more prone to this state of being than to mindfulness.

THE GOLAN HEIGHTS

'Not there, Chris. Mines!' shouts Yuval from the road.

I slowly retrace three steps. 'OK this way?' I ask, nodding to the left.

'Yes,' he says. So I go left.

The rocky hillside overlooks the silvery-still Sea of Galilee in the distance. The arched hill is spattered with blood-red anemones (*Anemone coronaria*) and skeletons of spent asphodels (*Asphodelus ramosus*). The land has a calmness to it, as if it was forgotten by people; a place quietly reclaimed by nature.

Giant fennel (*Ferula communis*) stems weave their way skyward out of the rock, as sturdy as telegraph poles and the size of small trees. I've always liked them. They are far more statuesque than their culinary relative. A dozen stand like apostles on the hillside; meanwhile juvenile specimens, yet without flowering scapes, have formed bottle-green fuzzy oases that spot the parched stubble. The low thicket around them is rough and spiny. One of the spiniest plants here is gundelia (*Gundelia tournefortii*), a great thistle-like creature with milky veins and formidable thorns splaying out over the rock. It looks like a crown-of-thorns starfish clamped to the cliff. Yuval tells me that the young shoots, leaves and flower-heads are often eaten in these parts, which seems remarkable, to look at the thing. Joining forces with the gundelia is another well-armoured plant, acanthus (*Acanthus syriacus*). As a garden plant in rainier climes it has an exuberant leafiness; here the spiny leaves are knit close to the ground, making its stout mauve and cream cones of flowers appear disproportionately large. A particularly exciting find is *Kickxia aegyptiaca*, a plant I've not encountered before. Its stems and leaves are grey-hairy and mealy, as if they were dipped in flour. Along one side of the zig-zagging stems are rows of cream-yellow flowers the shape of snapdragons (a distant relative) and the size of a fingernail; from the angle at which I'm peering at them, they look rather more like miniature frowning French bulldogs than they do dragons, only with long, arched tails.

On the crest of the hill we find what we came for: the black iris (*Iris atrofusca*).

A flower's beauty is defined by its colour, its pattern, its hue, isn't it? Take away light, the hue, and what you are left with is black – a void, a shadow. It's remarkable, then, that a black iris is so beautiful. It's like a negative against the pale grassy scrim; an afterimage fixed to my eyes from over-exposure, from not being able to look away. Mind altering. My camera flash brings light to the equation. It picks out purples and grey textures from the jet-black shadows, like crushed velvet. But the pupils, at the base of the petals, remain black. Around them a cobweb of tiny speckles and veins emerges, like those of the Nazareth iris, only sprayed over with black paint. It's sombre-looking and powerful. Transcendent.

WE PULL UP by a roadside verge. The map on my phone tells me we're by the 1949 Israeli–Syrian Armistice Line. The rocky green hummocks are netted with disused orange tracks and spirals of barbed wire. Fine grasses shimmer in the breeze. Poking out of the wire are the leafy spikes of an unusual species of weld (*Reseda alopecuros*), stockier than its familiar cousin the yellow weed (*Reseda luteola*) that was once used widely as a source of natural yellow dye. A wild birthwort (*Aristolochia bottae*), sprawling along the ground, is an exciting find. Its tubular, pitcher-like flowers have a flared, camel-coloured lip, blotched and marbled

Black irises (*Iris atrofusca*)
in the Golan Heights.

with Coca-Cola brown. They are fly-pollinated and they smell strongly of carrion. A few metres away, an impressive milk vetch (*Astragalus macrocarpus*) straddles the track, with long leaves floating upwards on an invisible current, and bunches of hairy, egg-like fruits.

We stop at Nazareth to see another population of its namesake iris. We join a long line of cars gleaming in the evening sunshine. I ask Yuval why the drivers are all tooting their horns, curious to know the object of their frustration. 'Chris. This is the Middle East. We do this just to say hello,' he tells me. Then he gives his own horn a long, hard blast, to which none of the other drivers pay the least bit of attention.

We comb the steep hillside on the side of the city, overlooking grey ribbons of land that repeat themselves and fade into nothing. The Nazareth irises we find have more of a mustard-yellow tint to their fall petals than those we saw this morning, and are torched gold by the evening sun. Their billowing standard petals look like discarded tissues strewn over the hillside. Some angry-looking youths wander over, apparently concerned that we're disturbing the plants, or picking them perhaps, I don't know. Yuval explains that we're scientists and that we're doing no harm, so they saunter away. Out of the intricate zigzags of thorny burnet (*Sarcopoterium spinosum*) I notice the slender green stems of *Cruciata articulata*, their ends arching upwards, as if held by invisible threads. Their lines of leafy bracts have a pleasing symmetry. They are like strings of hearts spilling out over the land.

THE JUDEAN DESERT

Today I'm joining Ori Fragman-Sapir on an expedition to the south of Israel. Ori is a good friend, a fellow plantsman and the Scientific Director of the Jerusalem Botanical Gardens, so I'm in good company. In his four-by-four we exchange stories about plants we have seen around the world as we hurtle south to Arad under a broad blue sky. We park on a stony parcel of land edged by the bleached skeletons of shrubs, where Dar Ben-Natan is waiting for us. Dar is a young botanist whose knowledge of where to find plants here is legion; there isn't a plant species he hasn't found in the Israeli deserts. We cover ourselves with various garments and creams to shield the fierce sun, and slip into the rocky ravine.

We're here to carry out a botanical survey in the Judean Desert. The region has enjoyed uncommonly high rainfall during the winter months that has rekindled life out of the rock. Dar has found alliums this spring not seen in flower here for 14 years. The sides of the tortuous ravine spooling out in front of us are brown and crinkled like wrapping paper. I run my hand along a splintered rocky shelf. It emanates its own warmth, as if by some geological process stirring beneath the surface; it feels alive. Which, actually, it is: the cracks and crevices that criss-cross its surface are glistening green, yellow and white with little desert annuals. It has a living sheen of flowers.

Ori trots off plant names like a machine gun. I note them all down in my pad. I know quite a few already, but I nod politely, or proffer names to those I'm familiar with, and he nods, signing them off. The first plant of note I stumble across is a creeping patch of ashwagandha (*Withania somnifera*), sometimes known as Indian ginseng. It has rather plain-looking grey-green leaves that obscure its inconspicuous little yellow flowers. Not the most prepossessing of plants, perhaps, but it is has long attracted attention, being prized for its perceived stress-relieving effects and used widely as a narcotic and sedative. Its Arabic name *saykarān* is believed to be derived from the Proto-Semitic for 'intoxicated' and 'inebriated', alluding to the plant's sleep-inducing properties.[20] Its uses were even known to the Ancient Egyptians, apparently. I've read that here in the Middle East its crushed leaves were once used as a poultice for sunburnt skin. I mentally note this, as the fierce midday sun sears the back of my neck.

Dar calls and ushers us over to see the allium (*Allium aschersonianum*) growing out of a pocket in the rock. It has a spear-straight, slender grey stem topped with a pompom of silky pink stars. Less than a metre away is a startlingly red tulip (*Tulipa systola*). The pinks and reds look almost synthetic against the thirsty brown rock. I pick my way down the rocky cliff to a sandy torrent bed. It's choked by a thicket of taily weed (*Ochradenus baccatus*), which is a common desert sub-shrub ornamented with upward-pointing yellow cones of flowers. Next to it stands a rare species of Jerusalem sage (*Phlomis platystegia*), with clusters of lemon-yellow flowers shaped like little frowning sceptre-heads. The sun casts jagged shadows over every leaf and flower, depicting them all in high definition. All things considered, it's strange that the word 'desert' suggests emptiness, a void. In just two hours in this smouldering ravine we've seen so many forms of life that we've only walked 30 metres between us. I wish I'd brought a bigger notepad.

On the way to the Dead Sea Valley we stop to observe another allium growing on a plat of wasteland. Wasteland doesn't sound promising plant-hunting territory, does it? Well, it is in this part of the world. The disturbance brings up all sorts of unexpected plants after winter rainfall. It reminds me of searching for sea creatures after a low tide: you never know what might turn up. We fan out over the biscuity ground with our eyes intent on the ground, pointing out various annuals along the way. The jointed, succulent twigs of *Anabasis articulata* form intricate tufted mounds around which sand collects, some bleached white like coral. Among them we find what we're looking for: *Allium rothii*, and it is exceptional. It has a stem as tall and thick as a pencil and blue-purple like a plum. Its small starry flowers are congested in a ball at the top, each with petals like baking paper, flushed claret and dominated by shiny black carpels and stamens. The overall effect is like a writhing scrum of glossy beetles. Black flowers really do have a beauty all their own.

The Judean Desert tilts into the Dead Sea Valley through a course of folding, honey-gold terraces. Its mudstone hills are devoid of trees and look naked against the clear blue sky; stripped to their bones they have a particular beauty. Viewed from above, the land would look like crumpled brown paper. We snake south towards a wrinkle called Nahal Zahav – a wadi where rain collects in the winter, creating an oasis. Other than the road, the place is untouched by civilisation; we don't see a building for miles. Ori off-roads the vehicle onto a rocky flat and we file purposefully into the pleated mudstone desert, shielding our eyes from the glare. Ori, Dar and I share the same love for desert floras and our excitement combined is like static. Dar and I dart about while Ori keeps a steadier pace behind us; together we clock dozens of interesting plants.

On the slope above us, sparse stands of khaki-coloured acacia trees (*Acacia raddiana*) hold their tiered branches aloft in the shimmering heat as if underwater. At our feet, puddles have blistered and cracked the earth into triangular, parchment-coloured shards, each as smooth as silk. Most of the vegetation will last for a just few short weeks. *Aaronsohnia factorovskyi* sprawls out of the crevices, bearing happy-looking flower-heads like little golden orbs. Some annuals, *Aizoon hispanicum* for example, are so small they could fit comfortably on my fingertip. Among the drifts of these more commonplace annuals we discover a rare jewel: a desert milkvetch (*Astragalus intercedens*) with pale pea-like flowers that point to the sky. A mound of white rubble is festooned with the withering vines of *Cucumis prophetarum*. Most of its leaves have turned to ash, but its spiky fruits are still ripening; they have the proportions of a lemon and the colour and markings of a watermelon, to which it's related. Nearby we discover the virile, pineapple-yellow eruptions of desert hyacinths (*Cistanche tubulosa*) (more on these later), pushing forcefully out of the earth. We stop to photograph the unignorable spectacle. For a moment I see us as others would: three men crouching in the desert with cameras pointing in opposite directions. You have to know when laugh at yourself, haven't you?

On the way back to Arad we stop at a desert plain north of the Bedouin town of Kuseife. Here on the stony turf stands a flock of irises (*Iris atrofusca*), shimmering like black taffeta under the softening late afternoon sun. Among them is a paler form with petals an undecided white bruised purple and ochre; together the dark and pale irises are a beautiful floral schizophrenia. On the other side of the road, an extravaganza of desert spring bulbs is quietly blossoming. A forest of black Persian lilies (*Fritillaria persica*) stands to a metre tall, like sentinels in the grass. Their grey leaves and dusky flowers are covered in a whitish waxy bloom. Like the irises over the road, there exists a pale morph among them; its fawn-coloured flowers blush ever so slightly red and hang like lanterns. In the understorey of the Persian lilies we see a clutch of ivory-coloured alliums (*Allium israeliticum*) pushing up out of the stubble, alongside a vigorous milkvetch (*Astragalus aleppicus*) tiered with spreading leaves like the branches of a miniature cedar tree. Perhaps most beautiful of all are the gladioli (*Gladiolus atroviolaceus*) with their wiry stems, to which bird-shaped flowers cling. They are the colour of clear night skies.

BACK IN TEL AVIV I have a beer with Eliran, the guy I'm staying with. Honeyed sunlight pours into his apartment through the open French doors. I show him photos of the plants I've captured on my phone today. He looks vaguely interested I suppose, mainly in the black iris. 'Sababa' – the Arabic word for 'cool' –

he murmurs, as he peers at my screen. He starts a conversation about film production. I find my mind drifting to irises and black Persian lilies, their pleasing symmetry and geometry. I make a mental tally of all the plants I've seen today while I pretend to listen to him. *Fourteen, fifteen* . . . Eliran is looking at me intently, his legs crossed and his left foot tapping gently in his grey flip-flop. I realise he's just asked me a question – something about music, I think. I make something up. As I reach for my bottle of beer on the table top, I notice a photographic collage of soldiers and guns frozen in the glass.

I take a taxi to the old city of Jaffa to meet Ori for dinner. The taxi driver is talkative. He likes to look me in the eye when talking which makes for an interesting journey. When I arrive, Ori and his wife Vered show me their apartment. It's a treasure trove of antiques and curiosities amassed from Ori's botanical expeditions around the world. He lists the artefacts and their origins like he does plants. His balcony tumbles with succulents escaping from their little terracotta pots. We leave the apartment and wander along orange-lit streets of stone, full of people, noise, and hipsters holding animated conversations through a blue cannabis fog. The city feels electric. Ori and Vered treat me to plates of creamy *masabacha* topped with chickpeas and olive oil, and pitta stuffed with aubergine, along with various salads, dips and a dish made with cauliflower so flavoursome that it changes my opinion of the vegetable forever.

THE WEST BANK

The desert is as bare and white as snow. It hurts to look at it. Drifts of ashen sand spill out on either side of the Al Bah Al Mayet road and extend as far as the eye can see. It's a bleached and empty landscape, strangely Arctic in appearance; a lifeless place. Except that it isn't. Because towards a desert spring called Ayn

Abu Mahmud (according to my map), an electric bolt of yellow catches my eye from the car. We – Ori, Dar and I – all know what it is. It's what we came for. We abandon the vehicle immediately and gravitate purposefully towards it from slightly divergent directions, like filings in a magnetic field.

The parchment-coloured earth is baked and has cracked into a jigsaw of convex jagged pieces, caked white by salt in places. It shatters under my feet. Few plants can grow out here in the dust. One that can is tree purslane (*Atriplex halimus*), a rangy shrub the colour of silver that lines the damper pockets of the desert. Another is its parasite, the desert hyacinth (*Cistanche tubulosa*). We soon find the great hunk of it that caught our eyes from the car, a yellow vision erupting from the earth. It has six ambitious stems that rise to my knee, as thick as saplings, jammed with flowers browning off around the base. We find a score shooting up from the bare ground, and they are an astonishing sight to behold. Like a mirage.

We set to work collecting specimens, pressing and dissecting flowers and dropping bits of tissue into tubes, each with a dollop of silica gel to preserve the plants' DNA. Our mission is to elucidate the species relationships among the desert hyacinths that occur here in the Middle East. These plants have been harvested in the region for use as famine food and herbal medicine, possibly since the ninth century; related species in China have been used for over 2,000 years.[21] Yet despite their intrigue and importance, little work has been carried out on their relatedness. We tend to think of species as fixed entities. But they change over space and time – they are four-dimensional. In plants such as desert hyacinths, which lack leaves and preserve poorly, understanding where one species ends and another begins can be complicated. Especially here, where the plants seem to have been all but ignored by botanists. So over the next few days we will comb the length and breadth of Israel and the Palestinian Territories in pursuit of desert hyacinths.

WE FAN OUT over an oatmeal slope perched over the Dead Sea.
A haze has formed around the mountains of Jordan, blotting the
lines of definition between sea, land and sky. The dazzling slab of
land sears itself into my eyes; I can still see it when I blink. It's a
shadeless place; hot, as dry as powder. And yet it teems with life.

Ori coaxes me towards a fold in the rock. 'What do you see?'
he asks, smiling expectantly. I peer at a puddle of orange dust and
see nothing. 'Look again,' he says, and I follow his eyes to the left.
I see a raft of wrinkly, succulent stems of a similar shade to the
rock. It looks like a mound of coral: one that drew its colour right
out of the earth. I crouch to examine it, my palms pressed to the
baking-hot ground. 'Caralluma sinaica!' I exclaim, to which Ori
nods, beaming. 'This is the finest specimen I know' he tells me. To
my delight, the lifeless-looking stems have burst into little fawn-
coloured starfish-shaped flowers. We take photos and I make a
quick sketch.

We clamber over a hill as naked as the moon, past lanky, dust-
coated stems of *Anvillea garcinii* rising out of crevices, clasping
the odd sunflower-like flower-head. The clayey deposits between
the rocks are dotted with rose of Jericho (*Pallenis hierochuntica*),
a form of 'resurrection plant'; this is a description given to a range
of plants that shrivel up in conditions of drought and are restored
by rainfall (the specimens in front of me are in full growth). Ori
tells me that *Periploca aphylla* can be found growing 20 metres
beneath where we're standing. He knows me well enough to
realise that this is my kind of plant. Like an excitable kid, I skip
down the steep, torrid slope towards a scratchy thicket. I spot it
immediately and decelerate abruptly to a halt. *Periploca aphylla* is

not the sort of plant that would normally stop you in your tracks: it's a spindly shrub, rather like an under-nourished tamarisk, holding stubbornly onto its dead branches and looking perpetually in need of a good drink (which is understandable here). Its wiry stems lean seaward, criss-crossing the horizon. It's the flowers that astonish. They are borne in little ball-like clusters clothed in white fur. Each flower is a twisted pentagon the size of a thumbnail with a claw of five crooked, black, spidery prongs. The overall impression is one of a cluster of inverted tarantulas, bound at the thorax, with their legs projecting.

Back at the top of the cliff, Ori and Dar are searching for something more elegant. It's a bad year for *Trichodesma boissieri*, Ori tells me, and we may not find it. The three of us braid the land with eyes fixed to the ground, accelerating every now and then when something promising catches our eye. Just as we agree to capitulate under the strengthening midday sun, Ori spots one growing out of a rocky furrow. We all peer at it. It has floury leaves and wiry stems that coil into tiers of downward pointing stars. The stars are the colour of opalescent skies, each smudged with caramel and stretched into a pointy white cone in the centre. We leave the West Bank via a military checkpoint zigzagged by silver fences, street lights and serious-faced soldiers garbed in khaki and parallel-slanting guns, who stand watching us.

THE ARAVA VALLEY

We park in a minefield. This is not a metaphor: I nearly slam the passenger door into the yellow 'DANGER MINES!' sign as I hop out. I casually ask if it's safe to walk here. 'Yes,' says Ori simply, so I follow his footsteps closely, out onto the stony plain in the soft evening sun.

The plastic sheeting of a nearby polytunnel crammed with pepper plants flaps and twists in the breeze. A clear plastic glove

A rare allium (*Allium sinaiticum*) in the Arava Valley.

glides by like a ghost hand, inflated by the wind. Ori marches us to the middle of the plain. We're here to see a rare allium (*Allium sinaiticum*) which grows along sandy flats beneath a rocky spine that straddles the Israeli–Jordan border. We soon find the plants sprouting all over the place. The unusual pattern of winter rainfall has persuaded them out of the earth in an abundance unseen in decades. They've been on strike for 20 years. Their short, stocky stems arch out of the ground and bristle with green-striped, white flowers. Yet another milk vetch (*Astragalus trimestris*) grows in the dust as well. Its shiny fruits are like a mass of writhing snakes. By the time we leave, the soles of our shoes are thickly carpeted in the thorny, button-like fruits of neurada (*Neurada procumbens*), which prove too painful to extricate by hand.

We spend the night at a ranch in a farming village close to the Jordan border. We eat a simple meal of cucumber, humus and various cheeses. At night I can't sleep. Tormenting neurada spines. Colliding thoughts. Thoughts of alliums, irises and unresolved desert hyacinths. Minefields. *Fullness of mind.* Vehicles thunder past my bedroom, spiriting away truckloads of peppers from the plastic furnaces to hungry European supermarkets.

In the morning we make our way bleary-eyed to the Sheizaf Nature Reserve. There are scarcely any roads so Ori off-roads the vehicle, taking us right into the very heart of the desert. The dunes are soft and rippled like a beach. The flats around them are woven with rivers of stones and torrent beds, each dotted with acacias like upturned green umbrellas. Here on the golden sand we find a rare tongue fern (*Ophioglossum polyphyllum*) unfurling. It's astonishing to see a leafy fern in the depths of the thirsty desert. Its leaves are reflexed and its curious spore-producing structures poke out from its crown. The whole plant could sit comfortably on my index fingertip. They look like the great welwitschias of Namibia in miniature. Like them, I suppose they must subsist on mist and dew – because in some years here, the rain never comes.

A desert hyacinth (*Cistanche violacea*) in the Arava Valley.

Paracaryum rugulosum
near the Sinai border.

EILAT TO THE SINAI BORDER

Stony shards glint in the sand like glass. Pillar-like date palms form neat, symmetrical rectangles of forest. Everything glitters under the hot sun. A petrol-blue haze blurs the horizon. After we pass Tsofar, the vegetation disappears except for the intricate skeletons of saltwort bushes that stud the flats under the fluted cliffs. The earth is brilliant white beneath the cloudless sky; in places it forms little pyramids as neat and fine as caster sugar. Everything is monochrome white.

Until we see purple. 'Stop!' we shout, and Ori crackles to a halt on the verge. Dar and I wander over to the desert hyacinth we have spotted. Its flowers are crisp and purple, like newly emerged butterflies. Its desolate surroundings make it look animated and somehow less plant than animal. I can imagine one punching its way through the sand in front of me in real time. This spectacular purple form of desert hyacinth has been recorded here before under the name *Cistanche salsa*. But we can see instantly that it has been misdiagnosed: *C. salsa* has hairy bracts (an important characteristic in these plants), and what is in front of us is as smooth as wax. We examine it closely, peering into the flowers, taping its bracts to sheets of brown paper, measuring and scribbling (the devil's in the detail). Then we christen it *Cistanche violacea*, a hairless purple species recorded frequently in nearby North Africa. But honestly, right now I couldn't care less what it should be named, or how we define it. Its beauty is just in its being: it is existential.

Chaos was the law of nature; Order was the dream of man.
Henry Adams, 1907.

Back at the car we find Ori busying himself with a patch of desert annuals. I think he may be tiring of desert hyacinths now. We zip down the main road towards Eilat, Israel's southernmost port on the Red Sea. We park on the outskirts of a white blocky resort by some rust-coloured pyramids of rock. In a gully we spot *Blepharis attenuata*. A violently thorny member of the acanthus family, it has upside-down looking blue-violet flowers poking out like panting tongues between the spines. Conditions are harsh down here. There is little else to see so we abandon the cliff and head west to the Sinai border.

An enormous tank rumbles past as we poke about the vegetation beneath the great metal border fence. A soldier watches us suspiciously from afar. He won't know that we're examining *Paracaryum rugulosum*, a distant relative of the forget-me-nots. He probably doesn't care. It has little plum, balloon-like fruits tethered to its stems. They look as though they could float away in a trice. It's surprisingly exuberant, as if someone has secretly been watering it out here in the desert. Or peed on it. 'Let's go,' announces Ori suddenly. 'People are edgy at borders.' We file into the car. But just as we set off, a soldier waves at us to stop. Ori opens the window to the passenger seat in which I'm sitting and leans over me to talk to him. The soldier does look edgy. He's wielding a large gun pointed at my arm. I slither into my seat. After a brief, animated exchange I can't understand, soldier and gun retreat, the window glides up again, and we're back on our way. I ask Ori about the situation; he waves it away with a slight brush of his hand. Dar tells me what kind of rifle he was holding and why it was so big. It sounds more dangerous than your average gun, but I don't know.

THE NEGEV DESERT

We stroll along the seabed. One that dried five hundred million years ago, only it looks as though the sea parted just last week. An empty, wind-beaten land of beige rocky hills, minted by the sea and baked dry by the sun.

Dar shows us the way. Somebody less attuned to the land might lose their bearings out here, for everything looks the same

A desert hyacinth
(*Cistanche fissa*) in
the Negev Desert.

in this remote tract of the Negev Desert. But not Dar. He leads us purposefully over rocky terraces bound by silty loess, past intricate mounds of *Haloxylon* splaying out like staghorn coral. A kilometre from the car, he tells us we have reached our destination. Here, on this rocky triangle, he's seen something unusual; and we can sense its presence, all three of us. The hairs on my arms prick up and my fingers fizzle in anticipation. Because there, thrusting out of the earth in front of us, is a desert hyacinth the likes of which I've not seen before. It has a fat cone of liver-coloured buds clothed in bracts, dusted in icing sugar. The lowermost flowers are just peeking open, displaying cream and raspberry petals. It looks good enough to eat. We peer at it, scribble down notes and leaf through photocopies of ancient texts on the genus that we've brought with us. Given its pattern of hairiness and bract geometry, the best fit for the plant is a species called *Cistanche fissa*. But *C. fissa* has never been recorded outside of the Caucasus east to the 'Stans. So the question is: are we looking at the first record of a species in this part of the world or a new species altogether? Nobody knows the answer to this. The answer lies encrypted in its DNA.

THE SUN HAS melted over Ashalim and gilded its bony, rain-scarred cliffs that pour out onto the verge. Like bone, these cliffs support living matter. We clambered up these rocky plinths to see a desert hyacinth (which won't surprise you), but now other things have caught our eye. Leafy crowns of *Astragalus caprinus* hug the bronze rock like green starfish. Clumps of desert campion (*Silene villosa*) burst into white flower like their lives depend on it, which they do. It's all going on here in the Negev.

Eminium spiculatum in the Negev Desert.

Iris mariae in the
Negev Desert.

The plant that excites me most on the sandy plateau is *Eminium spiculatum*. I won't pretend it's beautiful. Not in the conventional sense anyway. But it is *curious*, no one could argue with that. It's an aroid, rather like a ground-hugging form of the great dragon arums of Crete. Its spathe is like steak, mottled and brown, well done in the centre and rare around the edges. It has a matching spadix, only smoother and darker, like melting chocolate. And its leaves are like out-turned hands, pushing their way up from the sand. I have to tear myself away from its spell.

The rippled sand turns to copper foil in the fading light. These are my last few moments in the desert; I'm leaving tomorrow. But I don't want to think about that now. There is one more plant to see. Ori marches the three of us over the sand in creeping sunset, our shadows extending. I think he can sense I'm upset about leaving this place, and this is his gift.

A Negev iris (*Iris mariae*). A species I've not seen before. It's beautiful in every sense. Its three fall petals are frilly orbs of mulberry silk, intricately filigreed with darker veins. Each is brushed ebony at the base, like an ink stain spreading out along the creases. Above the falls arch a triad of petals the colour of twilight, folded neatly on top of one another like a pagoda. The flower shines like bronze and touches me deeply somewhere. But a gift like that is never free. I'm sad the whole way home.

In the beginning of March, the plains particularly betwixt Jaffa and Ramah were every where planted with a beautiful variety of fritillaries, tulips, and other pants of that and of different classes. But there are usually so many dangers and difficulties which attend a traveller through the Holy Land, that he is too much hastened to make many curious observations, or to collect the variety of plants, or the many other natural curiosities of that country.
Thomas Shaw, 1808.[22]

5

ON SACRED SLOPES

THE CANARY ISLANDS

5

ON SACRED SLOPES

THE CANARY ISLANDS

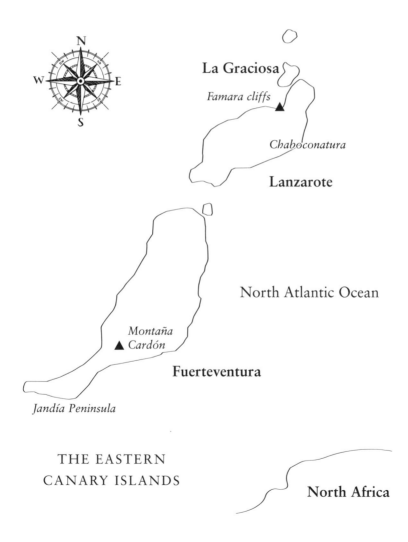

A bolt of electricity runs through me as I stand with the
giant succulents in the shadow of the Virgen del Tanquito:
her sacred slopes behind me, a land born of fire in front of me.

THE CANARY ISLAND archipelago has a famously diverse flora. This collection of Atlantic islands is home to 600 endemic species – plants found nowhere else on earth. Endemics account for 40 per cent of the islands' total native flora. The Canary Islands' exceptional plant richness is, in part, a result of their proximity to the African coast. Over geological time this acted as a bridge for species introductions and genetic exchange – processes often absent on more remote archipelagos.

Just 100 km (62 miles) from the coast of Africa lie the volcanic islands of Fuerteventura and Lanzarote, together with their drifting islets. These easternmost islands have a warm and windy climate classified as oceanic-desertic Mediterranean. Just 60 mm (2.4 inches) of rain falls per year around some coastal areas here, rising to about 200–250 mm (7.9–9.8 inches) in the highlands, with virtually none during summer months. They are as dry as dust, like fragments of the Sahara Desert, floating off into the Atlantic. In 1799 the German geographer and naturalist Alexander von Humboldt wrote:

> *The eastern islands, Lanzerota and Fortaventura, consist*
> *of extensive plains and mountains of little elevation;*
> *they have very few springs . . . The whole western part*
> *of Lanzerota, of which we had a near view, bears the*
> *appearance of a country recently overturned by volcanic*
> *eruptions. Everything is black, parched, and stripped of*
> *vegetable mould.*[23-24]

But Humboldt visited these islands in June. If he'd visited after heavy winter rainfall, he would have seen an astonishing

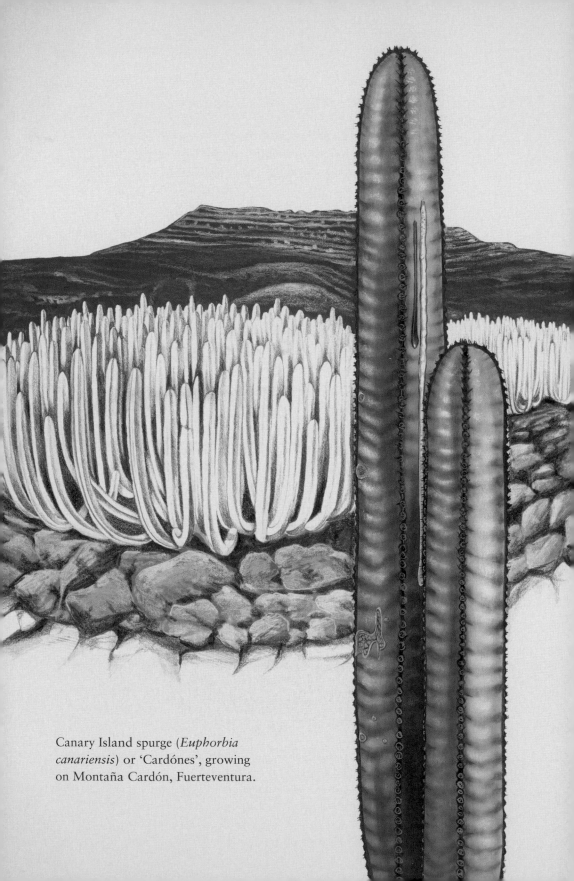

Canary Island spurge (*Euphorbia canariensis*) or 'Cardónes', growing on Montaña Cardón, Fuerteventura.

transformation: their stony orange plains and black rocky larva sheets would glow, not with embers, but with vegetation. Because for just a few weeks of the year, the islands become a mélange of green, yellow, pink and white; their crevices gleam; their roadsides flourish. For a botanist deprived of plants for several long, cold winter months, these warm oases of life are irresistible. For several years I escaped the cold milky skies of Britain to spend time with local botanists and ecologists, carrying out vegetation surveys and conservation work in the warm winter sun.

I've learned you can tell the weather on these desert islands by its plants; you can see when it has rained and how much. No two winters are the same. The rain falls in a different pattern each year, teasing plants out of the rock one winter but not the next. In some winters it scarcely falls at all – you need to search harder then to find the plants. That's when I've chased them down cliffs and cornered them on death-defying vertical black walls that bathe in the mist, walls as high as the tallest buildings in the world. It's enough to make my palms go clammy even now.

FUERTEVENTURA

A sunny day in December. I head south to the Jandía Peninsula under a broad, blue sky to search for a special plant that grows only in this empty corner of the planet. The deserted road weaves its way across bumpy orange plains studded with white windmills and lonely goat farms. I pass a small village named Tuineje and, to my delight, spot desert gourds sprawling over a stony flat. Their green and yellow marbled orbs are strewn about like an upset apple cart. I park the car and wander over. They look like a field of watermelons in miniature. I pick one up. It feels satisfyingly heavy in the palm of my hand. I make another stop at Cardón, a quiet village nestled at the foot of the great Montaña Cardón. Both town and mountain take their name from the very plant I plan

Montaña Cardón,
Fuerteventura.

to see: the great Canary Island spurge (*Euphorbia canariensis*) or 'Cardón', a plant once much more frequent and harvested historically by islanders for fuel. It clung on to just a few remote slopes on this fast-changing island, which makes the prospect of finding it very special.

The town is eerily devoid of people. Collared doves chant 'koo-*koo*-koo, koo-*koo*-koo,' from palms that make a plasticky sound as they quiver in the breeze. I head to the foot of the mountain. It looks like a lazy great rhinoceros, sprawling over the horizon. Pleated mounds of yellow and orange earth pour out from under its feet. Montaña Cardón is the most spectacular of the mountainous massifs on the southern mainland and the island's richest floristic region after Jandía, with an estimated 18 endemic plants. The mountain's steep and imposing rocky walls are incised with narrow valleys and damp crevices shielded from human impact. It's steeped in legends of heroic warriors and today is a place of pilgrimage and worship for local parishioners, who dance to their Virgen del Tanquito. It's 690 m (2,264 ft) high – modest really – but the great reddish-grey mass feels larger, jutting out of the flat Cofete Plain. Deep beneath the surface, from its core, spreads an unseen labyrinth of remnant volcano ducts through which larva once spewed.

Approached from the west, a well-trodden track leads to the mountain's summit. But the great stands of Cardónes I'm after grow only on the precipitous crags of the mountain's eastern shoulder, to which there is no path. I'll need to make my own. I scan the landscape for the most accessible route up and set off. Green and orange fields form a patchwork over the plain, but the mountain's slopes are barren. I pick my way over the lunar landscape. It's an empty place but for the sparse grey spiny balls of *aulaga* (*Launea arborescens*), *espino* (*Licium intricatum*) and saltwort or *mato* (*Salsola vermiculata*), all of which are common here. Most are leafless, but the saltworts are still hanging on to last season's carmine-coloured fruits. Fat, hand-like *verodes*

The 'Jandía thistle' (*Euphorbia handiensis*)
growing in Jandía, Fuerteventura.

(*Kleinia neriifolia*) grope at my ankles out of the rock. The slope is incised by cavernous gullies carved from torrents over the course of millennia. The largest of these is the Barranco de Rincones, and I need to clamber my way down, into and up out of it to reach the red haunches of the mountain. I can see the Cardónes lurking there, in the distance.

Breaking a sweat, I heave myself up onto the slope. Suddenly I am surrounded by them: enormous stands of fat, yellowish-grey organ pipes rising several metres out of the bare rock, with the summit of their namesake mountain looming above. Cardónes are everywhere: a hundred or so, growing in ranks along the barrancos that tumble down the mountainside. Growing out of the naked rock like this, they remind me of giant sea creatures – deep sea coral, or tube worms perhaps, thrusting out of the ocean floor. Their vertical columns are scarred and blackened at the base; it's hard to estimate their age, but I guess they may be centuries old. I feel a bolt of electricity running through me while I stand with the giant succulents in the shadow of the Virgen del Tanquito: her sacred slopes behind me, a land born of fire in front of me.

I continue south to the Jandía Peninsula, Fuerteventura's botanical hotspot. Once an isolated island, Jandía is now a rocky appendix to the main island, connected by a broad, white sandy isthmus, made up of millions of years' worth of shells, sea urchins and algae. The great grey vertical prisms that delineate the long, volcanic peninsula loom up over the horizon. These steep and inaccessible cliffs have been sheltered from human impact and grazing, and so they are home to an interesting summit-scrub vegetation, including plants that are absent elsewhere on the island. But the plant I'm after grows at the very foot of the cliffs. It's Fuerteventura's most celebrated endemic: the 'Jandía thistle' (*Euphorbia handiensis*). Like the Cardón which I've just seen, this iconic plant is a succulent euphorbia. It's confined to just a couple of remote rocky valleys in the Parque Natural de Jandía.

After a string of a tourist resorts, the road peters out into the dust. Mid-December and the landscape is completely parched. It looks as though it hasn't seen a drop of rain in years. Lazy blue sea glitters around the thirsty brown hills. I slip into the barranco and turn right, sensing the plant before I see it. It isn't long before it comes into sight. Just minutes after leaving the car, I spot the spiny grey stems writhing about the boulders like a Medusa's head. I kneel to look at one. Its columnar stems are interlocked, forming a domed, thorny cushion. Its spines look ferocious, like those of a porcupine. They give the plant a malign appearance, a sort of intentionality, as if it means to inflict harm. I wander over to the next one and realise there are several dozen nestled among the boulders of the ravine. In a remarkable feat of convergent evolution, this plant bears an extraordinary likeness to the unrelated cacti of the Americas: a resemblance shaped by the selection pressures they have shared in the harsh environments they made their own. I settle next to one and sketch it in the warm breeze, forgetting about the world for a while.

The current drew us toward the coast more rapidly than we wished. As we advanced, we discovered at first the island of Fortaventure famous for the great number of camels which it feeds; and a short time after we saw the small island of Lobos in the channel which separates Fortaventura from Lancerote.
Alexander von Humboldt, 1799.[25]

LANZAROTE

Lanzarote, which lies to the east of the archipelago, is one of the best-known Canary Islands, owing to its accessibility and gentle climate. Holidaymakers flock here for its warm winter sun, lunar landscapes of jagged volcanic peaks and white, sandy beaches. Far from the bustling holiday resorts in the south of the island, the Famara massif in the north-west is a geologically old formation with high cliffs and ridges which remain inaccessible to people and goats alike. Most of the island's endemic plants are concentrated here. Twelve are found only here, in fact. These black folded peaks rising out of the Atlantic mist are my destination this winter. I've planned botanical excursions with the local botanist, Alfredo Reyes-Betancort, who is the regional authority on the Canary Islands' flora. He grew up on Lanzarote and knows every plant, every cliff and every foothold on the island.

Our excursion begins with a gentle clamber down the jagged slopes of Las Rositas, which pitch calmly into the Atlantic. We walk through a forest of giant, yellow-flowered sow thistles (*Sonchus pinnatifidus*) to the edge of the cliff. Alfredo points out prickly arching branches of *esparraguera majorera* – an endemic wild asparagus (*Asparagus purpuriensis*). A tough, thorny mass, it seems a far cry from its edible cousin. Here and there, wild white alliums (*Allium canariense*) are peeking out of the rock and wild lavender (*Lavandula pinnata*) forms violet splashes by the path. It's been a dry winter and the spring-flowering annuals are patchy. We'll have to search hard to find them this year.

We continue down the awkward track until we reach the *locus classicus* (the first recorded location) of a rare rockrose (*Helianthemum thymiphyllum*), which grows on just a few rocky shelves overlooking the sea here. Following the dry winter, few are in flower. Some have scarcely any leaves. We scrutinise the plant's features, its hairless, deep green shiny leaves and brown bark – all the diagnostics of the species. Back at the top of the cliffs we hunt for a broomrape (*Orobanche castellana*) which, needless to say, I'm excited about. We scratch about for 15 minutes, but sadly none are in flower. So we drive south, in parallel to the Famara cliffs, which form a crest along the north of the island. We pass the quaint hill town of Haría, set between the peaks of Peñas del Chache (known as 'the valley of a thousand palms'), where, to my excitement, giant fennels (*Ferula lancerottensis*) are erupting out of the rock. They look impossibly verdant and fresh on these desperately dry slopes. Their weighty taproots must probe deep into the fissures of the rock for what little moisture is retained there, fuelling the fat flowering stems they send up in winter. Nearby, high on the windy Mirador de Haría, we see a rare daisy-relative *Argyranthemum maderense*, the 'Madeira marguerite', which is cultivated far and wide for its lemon-yellow flower-heads. Alfredo wanders off to take a phone call and I sit on a rocky shelf for a moment, taking in the views over the island unfolding in front of me in the warm breeze.

As we drive south, I watch cliffs pour into the ocean in a frozen flow of rocky scree. We take a path that snakes round giant, greenish-black boulders perched above an amber plain. Rusting skeletons of fallen vehicles scatter the slope beneath us. Alfredo shows me his favourite endemic here, a curry plant (*Helichrysum gossypinum*), known locally as *yesquera amarilla o algodonera*. It means 'yellow tinder', he explains. The shrublet's fat grey cushions sprout out of the rocky crevices. I stroke its leaves and it is soaking, saturated with sea mist. Nearby we spot bronze, succulent cushions (*Aichryson tortuosum* and *Monanthes laxiflora*) swelling out of the crevices. We return to the track where Alfredo points out buttercups (*Ranunculus cortusifolius*) in full bloom, like puddles of sunshine. After a 20-minute walk, we clamber down a steep slope to see the scillas (*Scilla latifolia*). About a dozen leafy bulbs have fat stems, bulging with fruits. Alfredo shows me photos of the plants in flower, which he took a few weeks ago. On our way back to the car, a blanket of cloud blots out the sun, draining the land of its colour. Just metres from the vehicle, Alfredo points out a purple sand crocus (*Romulea columnae*) at our feet which we must have missed. It's about the size of a cherry. We crouch to take a closer look and just at that moment, a beam of sunlight falls on it like a spotlight. It looks like an amethyst, gleaming out of the rock.

> *The black mountains . . . appeared like perpendicular walls*
> *of five or six hundred feet. Their shadows, thrown over the*
> *surface of the ocean, gave a gloomy aspect to the scenery.*
> *Rocks of basalt, emerged from the bosom of the water,*
> *wore the resemblance of the ruins of some vast edifice;*
> *their existence carried our thoughts back to the remote*
> *period when submarine volcanoes gave birth to new*
> *islands, or rent the continents asunder.*
> Alexander von Humboldt, 1799.[26]

WE HEAD SOUTH to the amber and black volcanic plains on the centre of the island to see the endemic curry plant (*Helichrysum monogynum*). Unbefitting for such a special plant, we find it growing beside the main road with billowing white shopping bags snagged to its branches. Passing drivers stare as we kneel to examine it. Alfredo seems oblivious – perhaps like me he's used to that. We find great ash-coloured mounds of the shrub, a handful of which have scarlet flowers peeking through the intricate branches, like embers.

Back on the north of the island, at Famara, drifts of white sand have blown onto the road and sprinkled the tarmac like sugar. Sleek black surfers huddle on the sand and pepper the sea; motionless, waiting for the breakers. They look cold. We hobble awkwardly past them over the smooth grey boulders. Along the way, Alfredo points out rare plants that bathe in the ozone. It smells of salt and seaweed, and my skin becomes rough with brine. Gangly sprigs of *Periploca laevigata* are bouncing over the rocks. This is a plant I know from time spent in deserts; wherever I see it, its starfish-like flowers and antler-like fruits give me a thrill.

At the far end of the beach we scramble up a steep escarpment in pursuit of one the island's rarest plants, a miniature 'tree plantain' (*Plantago arborescens*). We search for an hour without success. Up and up we go, with our necks craned, scanning the cliffs above. Eventually, the palms of our hands cut by knife-sharp rock, we find a single specimen sprouting out of a gulley. I have to say, it's an unprepossessing thing. Regardless, we spend some time admiring it, given the effort we've put into tracking it down. On the way back down, we see white-flowered daisies *tojia blanca* (*Asteriscus schultzii*) stuffing the rocky fissures. Best of all, before returning to the car, Alfredo points out *piñamar mayor* (*Atractylis arbuscula*). The little domed shrubs are sprinkled with white flower-heads, like stars. The plant is threatened with extinction and grows only among these boulders in the thunder and hiss of the Atlantic. It's rarer than a blue whale.

> *The volcanic summits of Lanzerota, of Fortaventura . . . were*
> *like rocks amidst this vast sea of vapors, and their black tints*
> *were in fine contrast with the whiteness of the clouds.*
> Alexander von Humboldt, 1799.[27]

THERE ARE STILL a handful of plants left to see, but they won't be easy to find, explains Alfredo in the car. We pass jaded cyclists up the steep incline to El Bosquecillo, a 'miniature forest' perched on the island's summit. We leave the car by a gathering of walkers and piles of backpacks and walk to the edge of the vast *riscos* – the great black curtain of rock with buttresses that drop steeply to the sea. I watch a concertina of unbroken waves roll silently in slow motion towards the horseshoe-shaped bay below. Hikers stand gazing at the view, their clothes billowing in the wind. Leaving the footpath, we head towards the edge of the cliff. Alfredo asks if I'm comfortable with heights. I smile and tell him I am.

Imagine crawling out of the window of a 182-storey building. Hang on to the window frame with both hands, keeping your back against the wall, with your legs dangling in mid-air. Now, slowly, look down between your feet. Horrifying, isn't it? Horrifying but *thrilling*.

A strip of white sand stares up at me through the mist, 600 m (1,968 ft) below. The palms of my hands sweat. We inch down, with our backs to the cliff, pinned by slamming gusts of wind. Tapping the rock with the backs of our feet, we find the footholds and grip the cliff with both hands. Slowly does it. My heart pounds, my fingers tingle, and a cold sweat creeps along my spine. I know these feelings well: feelings brought on by a death-defying drop beneath my feet. All that air beneath me, yes the void – and the electricity too – that little frisson of excitement that comes with chasing down a plant. *Steady.* I fix myself into position with one hand to reach for my notebook and pencil and then . . . WHOOSH! A sudden gust slams me back against the rock. My pencil slips between my clammy fingers. I watch it bounce and fall for half a kilometre, then vanish into the abyss. The hairs on the backs of my arms prickle. To my right Alfredo, oblivious, is busy positioning himself next to the ledge. I clamber carefully down towards him. He looks up and smiles proudly. Then he reveals a veritable rock garden on the mist-soaked shelf of the

Descending the Riscos de Famara, Lanzarote.

cliff. I peer at the assortment of glistening wet plants. He shouts a gale of scientific names through the roaring wind. '*SIDERITIS PUMILA*,' he yells – this is the rarest in the bouquet. It's a gnarled, unenthusiastic-looking relative of the mints and lavenders, stunted by the wind-blasted conditions. A humble-looking thing, I have to say, but knowing I'm sharing a moment with it, up here in the only nook of this planet where it grows, tucked away out of sight: that makes it special. The shrieking wind picks up again. 'WE NEED TO GO,' shouts Alfredo. So with our cheeks to the rock, we leave the secret garden on the cliff and haul ourselves back over the precipice to safety.

The wind drops as we stroll along the footpath, and everything is calm again. I can't quite believe what just happened. 'Now you have seen every plant on Lanzarote,' says Alfredo, beaming, as we return to the car, with my hands and feet bleeding.

TODAY I'M MEETING the island's ecologist, Matías Hernandez Gonzalez, in the centre of the island. On the way, I notice a fresh green mantle has crept over the hills following the recent rainfall. It looks more like northern England's Peak District than it does a desert near the Sahara. The vegetation smells herby, like marijuana. I weave through green and black fields of stunted vines, sheltered by semicircular walls of volcanic rock and rows of sweet potatoes. Bent figures are farming the little fields, coaxing crops out of the ash with what little rain falls from the sky. Behind them extinct volcanoes look out on an audience of *Aloe vera* plantations. The fields give way to malpais studded with blue-grey mounds of succulent *Euphorbia balsamifera* and *Kleinia nerifolia*. Their finger-like stems poking out of the rock look like coral. I feel as though I'm driving through the seabed.

I find Matías waiting for me at the bus stop. As we wander along the lane, he tells me he inherited a family plot here in the village, which he's converted into a nature reserve to grow rare plants. *Chaboconatura*, he calls it. The hard-cut landscape of ancient larva is sharply defined in the morning sun. In places it looks as though it spewed out of a volcano just last month. We saunter around the warm black slope for half an hour, choosing specimens for me to paint. A guild of tiny flowering annuals has sprung up along the crevices, criss-crossing the black, lumpy ground with green. We identify a dozen species growing in just a few square metres, including little pink-flowered crane's bills and orange calendulas. The *Helichrysum monogynum* I encountered with Alfredo a few days ago by the roadside grows wild here, along with succulent forests of muscular-looking tree houseleeks (*Aeonium lancerottense*). Matías points out heirloom succulents planted long ago by his ancestors. Every so often we crouch to identify a little annual or examine the stipules of a medicago. He looks at me with intent dark eyes as I list the plants' names, nods and then scribbles them all down in his notepad.

In the afternoon, I paint the reserve's plants while Matías potters about in the shed making plant labels. My workstation is a rustic little wooden table and chairs set in the dappled shade of a California pepper tree (*Schinus molle*). Next to me, great columns of an ancient Cardón (*Euphorbia canariensis*) are forcing their way through the rock. A warm pulse of wind rustles the papery leaves, then skitters off into the distance. I lay everything out in front of me. My box of watercolour paints has a comforting aroma, like nutmeg. I arrange each of the leaves and branches carefully, shielding them from the wind with my hands. I look closely at their intricacies – consulting every scar and bruise that will add character and bring them to life on the page. Next I sketch the composition. Wildflower assortments can be tricky. Not all species are conventionally beautiful, but each has its own rhythm, and together, they must find harmony. Once I'm satisfied with the balance, the geometry, I soak the empty shapes with water. Then I daub them with carmine, vermilion and burnt umber. Clouds of colour sing from the page.

Oil paints are my medium of choice (the illustrations in this book are all oil paintings), but watercolours are portable and dry quickly, so they are perfect for painting *en plein air*. I have to adjust to painting in reverse though. I need to identify the palest areas first with watercolours, so they shine through at the end; they can't be glazed in later, as they are with oils. As I watch the pinks and yellows bleed into one another, little reptiles dart in and out of the rocks, doves coo from the canopy and bees buzz purposefully about the *Schinus* blossom. An Atlantic lizard (*Gallotia atlantica*) the length of my forearm shoots me a disapproving sideways glance, then vanishes into the wall. By the afternoon I've built the body of my subjects sufficiently to add the finer details: the leaf margins, veins and flower parts. It's satisfying to sharpen up the edges and give each plant its personality. As the sun falls out of the sky, I'm reluctant to pack up and leave this little volcanic oasis. But I feel lucky to have spent time in it, getting to know it, soaking up its colours.

IN THE EVENING I join a community-led effort to 'rewild' a derelict site in the city of Arrecife. The plot is owned by the city council, and is overlooked by the extinct Maneje volcano. Matías tells me it's been devoid of vegetation for 30 years and so is the perfect place to establish an urban community garden. Our aim is to establish native plants to create a green space that will benefit people's well-being and engage the community with the beauty and importance of Lanzarote's local flora. Matías looks proud when he tells me this – I can see it means a lot to him.

Everyone has gathered in the centre of the plot around a heap of rusty gardening tools and recycled plastic half-bottles, each containing a seedling nurtured by Matías on his balcony. I'm introduced to Javi, Matías's old school friend, and half a dozen families who live nearby and have come to help. Twenty of us fan out over the empty rectangle of land. It's carpeted in volcanic gravel freckled with white heliotrope (*Heliotropium bacciferum*). The flowers smell of paper-white narcissus and cat pee. One seedling we're planting is the iconic Cardón (*Euphorbia canariensis*) – the plant I saw growing wild on the mountain in Fuerteventura. Each one has a single purple, finger-like stem. It's hard to believe that it will achieve monstrous proportions one day. We also plant wild lavender *matorriscos* (*Lavandula pinnata*) and *verodes* (*Kleinia neriifolia*), both of which are common across the island. As our trowels chink against the rock, I realise that what we're doing feels a lot like tree-planting: fulfilling that desire to make a mark on the land; hoping that the plants will be here long after we're gone. One of the kids runs around excitedly, while Matías and Javi weed the

perimeter of the plot, pulling up invasive plants, and stuffing them into purple bin-liners that they then toss over their backs like a Santa Claus's sack. After an hour, the sun fades, and we all stand back to look at what we have achieved together. There's a quiet sense of satisfaction. Satisfaction that we've planted the seeds of a slightly greener future here in the city tonight.

Javi and Matías removing invasive weeds in Arrecife, Lanzarote.

LA GRACIOSA

I meet Matías at the little seaside town of Órzola where we will take a boat to La Graciosa. Known as 'the eighth Canary Island', Graciosa is one of a cluster of small volcanic islands that makes up the Chinijo Archipelago to the north of Lanzarote. Our quest is to locate the Canary Islands' only remaining population of 'Maltese fungus' (*Cynomorium coccineum*), which is the focus of a conservation project we're carrying out with the local authorities. Not a fungus in fact, *Cynomorium* is a parasitic plant. It's an extraordinary-looking plant that pushes up black spikes of flowers out of the sand like stalagmites. Its python-thick stems were once dug up as an 'emergency food' during famine. I've encountered this botanical oddity in the Mediterranean, but I've not seen it growing this far west, where it is poorly documented. Matías and I are curious to see it growing here and to assess its conservation status.

We leave Lanzarote in the shadow of the great grey walls of the Famara massif, the sea lashing its foot. We cross the Strait of El Río in a 20-minute gulp of sea spray and petrol, and watch a jumble of white houses framed by volcanoes come steadily into focus. As we disembark, the residents of Caleta de Sebo are busy preparing for Christmas. Bulging cacti and succulents have been wrapped in tinsel and baubles. I spot a plastic baby Jesus wedged into the candelabra stems of a succulent euphorbia tree. We stroll across the sandy road towards the first patch of vegetation we see, all fresh and dewy from the rain. A healthy clump of sea rocket (*Cakile maritima*) sprawls out of a pile of rusty blue metal drums. We crouch to scrutinise the leaves of a silvery shrub we identify as *Atriplex glauca*. Nearby, we spot a succulent clump of *Astydamia latifolia,* a relative of the rock samphire. Its thick, glossy leaves look incongruously fresh against the parched ground. I crush a fruit; it smells just like parsley.

We leave the small white and green fishing village of Caleta de Sebo and head east along a track skirting the Agujas Grandes volcano. A launea bush has been engulfed by writhing yellow dodder stems (*Cuscuta approximata*), like spaghetti. We make our way haltingly, identifying the plants we find and admiring the beautiful rocky coves. 'This one is called the Bay of Rabbits,' Matías says; then, laughing, he tells me the Lanzaroteños (people of Lanzarote) are referred to as Los Conejeros – 'the rabbiters'. He describes his adventures on the nearby island of Alegranza, the fist of rock we can see rising out of the ocean a few miles away. A rare colony of shearwaters breeds there.

We reach the spot where our plant is reported to grow. It's a sandy plain dotted with saltworts and the skeletons of dead *Launea* bushes. We dump our bags and comb the area forensically, our eyes fixed to the sand. We hunt for hours in the hot sun. A shimmering petrol haze blurs the horizon. 'There!' we shout, only to realise we're looking at the blackened stump of deceased bush. Rocks and branches trick us too, like mirages in the desert. *Cynomorium* is nowhere to be seen. Eventually we abandon our mission for fear of missing the last boat home. Disappointed, we walk the dusty track back to the port. We reach Caleta de Sebo in time to crack open a beer and quietly watch the sun set over the island.

It's a pity we didn't find it after all the anticipation. Everyone on the boat looks tired; Matías seems fast asleep. But I feel restless as I stare out to sea, as if something has been left unticked: a bed left unmade, a birthday forgotten. A painting not started. But at least I have a reason to return. I watch as one island fades behind me, and another rises up in front.

> *The land, which we had considered as a prolongation of the coasts of Lanzerota, was the small island of Graciosa . . . We took advantage of the boat to survey the land, which enclosed a large bay. No language can express the emotion, which a naturalist feels, when he touches for the first time a land that is not European. The attention is fixed on so great a number of objects that he can scarcely define the impression he receives. At every, step he thinks he discovers some new production.*
> Alexander von Humboldt, 1799.[28]

6
ACROSS KINGDOMS

JAPAN

6

ACROSS KINGDOMS

JAPAN

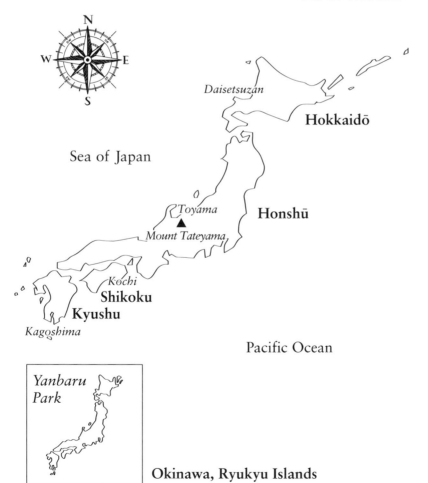

Sea of Okhotsk

N
W E
S

Daisetsuzan

Hokkaidō

Sea of Japan

Toyama
▲
Mount Tateyama

Honshū

Kochi
Shikoku
Kyushu
Kagoshima

Pacific Ocean

Yanbaru Park

Okinawa, Ryukyu Islands

It's difficult to make sense of how much natural beauty
we have seen in a short space of time.

TAKE A MAP of Japan. See that tilted stingray at the top? That's
Hokkaidō. Follow Wakkanai at the tip of its wing down to the city
of Asahikawa, just left of centre on the island. Now drop down
a little to Biei – a pleasant little town set in a patchwork of fields
– and pan right. That green blob? All 2,267 square kilometres
(875 square miles) of it – an area 40 times the size of Manhattan
– that is the Daisetsuzan. Keep going, right into the heart of it,
to where rivers replace roads and bears prowl uninterrupted by
people. Stop. There. Zoom in and that's where you'll find us:
four botanists, like ants in a line, heading deeper and deeper into
Japan's greatest wilderness.

Japan is one of 36 'biodiversity hotspots' – places that cover
just two per cent of the Earth's land surface, yet contain over half
of its endemic plants. Yes, the world's eleventh-most populous
country is home to a bewilderingly rich flora, with an estimated
6,000 different species. How can this be? Well, Japan's long string
of islands crosses two kingdoms: the Boreal Kingdom in the north,
and the Paleotropical Kingdom in the south. Plants bathing in the
warm, wet rainforests of Japan's southern islands have affinities
with those growing in South-East Asia; they are quite different
to those at the opposite end of the country, which are more akin
to those found in the frozen forests of Siberia. Take the botanical
legacy of Japan's former geological connection to the mainland,
throw in 15 million years of isolation and a range of altitudes and
climatic conditions – the odd monsoon and volcano included –
and bam! You have a crucible for the evolution of an exceptionally
diverse flora.

Japan's natural beauty fascinated early nature writers and
botanists the world over. Carl Peter Thunberg (1743–1828) was
a Swedish naturalist who, when stationed in Japan, traded his

medical knowledge for plant specimens with local interpreters.[29] He is cited in the naming of many plants, perhaps most notably the tropical genus *Thunbergia*. Philipp Franz Balthasar von Siebold (1796–1866) was a German botanist; he was also an avid collector of Japanese plants who established his own botanic garden in which to grow them. He is said to have introduced hostas to European gardens.[30] Following in their footsteps, Carl Johann Maximowicz (1827–1891) was a Russian botanist who also named many plant species in Japan.[31] The English clergyman, writer and mountaineer Walter Weston (1861–1940) was famously captivated by Japanese landscapes; he brought the Japanese Alps to the attention of the wider world with the books he wrote, for example *The Playground of the Far* East.[32] Both Thunberg and Weston wrote lively accounts of their expeditions.

My own glimpses of Japan, its plants and its people stem from participation in a decade-long campaign of conservation work carried out by Oxford Botanic Garden and Arboretum alongside local botanists. It aims to bank the seeds of rare species, collect plants for living collections and herbaria, and develop a method for measuring the species richness of plant communities. Let's pause for a moment to consider this last point: *species richness* – sounds good, but what does it mean in practice? Imagine a segment of forest – one you could hike through in 25 minutes: how many different species of plant are growing in it? Ten, twenty, fifty? Let's say 120 (a number plucked from a forest in northern Honshū). Now, let's consider *which* species are growing there – are any of them invasive? Or endemic? On the verge of extinction, even? What value can we place on each plant and what is their value as a community, collectively? Each of these things is important when considering how to prioritise areas for conservation in a fast-changing world with a growing population.

Conservation of a species-rich forest – it's obvious why that's important, but why press and dry plants in a herbarium? For centuries, botanists have done this, creating archives to document the world's flora one dried plant at a time. Traditionally these collections were amassed for curiosity, for documentation's own sake. Today, herbaria are becoming recognised as important repositories that can help scientists make sense of how our vegetation is changing because of habitat destruction, climate change and extinction. Historically, botanists may have collected specimens of plants they found interesting at a particular time and place. Our vegetation surveys in Japan amass samples from *all* the plants in a community – large, small, common or rare; this way we can track how whole plant communities change over time.

In this chapter I'm accompanied by Ben Jones, Curator of the University of Oxford's Harcourt Arboretum. Together we've trekked across Japan's 3,000 km (1,864 mile) string of islands from the Palaearctic mountains of Hokkaidō in the north, through the broadleaf forests of central Honshū and Shikoku below that, then down into the steamy, subtropical rainforests of the Ryukyu Islands. Hacking up mountains, sinking in swamps, dodging volcanic eruptions and fleeing typhoons, we've penetrated the very heart of Japan's wild unknown, crossing its kingdoms in pursuit of plants.

HOKKAIDŌ 北海道

'If it hears you coming,' explains our host Sakaue-san, 'the bear will not attack.'

We smile and nod – this sounds reasonable.

'But if it's surprised,' he continues, 'it will kill you.'

Ah. 'Have there been bear attacks here before?' I ask breezily.

'Yes, yes,' he mutters casually, batting away a cloud of insects with his hand. So we fasten bells to our waists, to rule out surprise, pull the mosquito nets over our heads to rule out the inevitable, and clang up the mountain track.

Today we're heading into the very heart of the Daisetsuzan (大雪山), Hokkaidō's largest national park – a rugged wilderness larger than some prefectures. We're surveying Mount Dairoku (大麓山), a remote lonely peak yet to brush fingers with civilisation. Sakaue-san and Kimura-san from the University of Tokyo lead the way. Sakaue-san is tall and serious; he has an air of authority – someone who leads Cub Scout adventures, that sort of thing. Kimura-san is a man of few words and a calm manner; I feel he connects more with nature than he does people. But we share the universal language of plants. And of the hundreds we see, there isn't one he can't name. Both Sakaue-san and Kimura-san are dressed in what looks like military gear, complete with weaponry fastened to their belts, and hard hats. The summit trail is all but enclosed by bamboo which Kimura-san hacks away with his machete. Swishing and slashing, I'm not convinced they aren't creating a whole new trail altogether.

Mount Dairoku is 1,459 m (4,787 ft) high, and the inhospitable thicket makes the climb hard work in places, so we pause regularly. Beyond a cloak of papery bamboo the view swoops down over a green and black mosaic of conifers and deciduous trees, up to a concertina of blue mountain cut-outs that fade into sky. As we approach the summit, the trail disappears completely, and we push through the branches that snag our mosquito nets. We clamber up splintered shelves of rock and finally emerge onto the great roof of the mountain. A boundless, hazy blue panorama unfolds around us, with not a road or building in sight. We admire the wide sky and snow-streaked blue ridges for some time, then stage a photograph, balancing our phones on a rocky ledge. Then we dump our bags and set to work collecting and identifying the plants we find among the stunted Sakhalin spruce (*Picea glehnii*) in the stiff, cool breeze.

Thankfully our descent is a more leisurely affair (now that we have a path) and we admire the many mountain plants we seemed to miss as we hacked our way up. Shinleaf or *Ichiyaku-So* (*Pyrola japonica*) is flowering along the track. The plant is used to staunch bleeding, say our hosts. The clusters of white bells look like pearls strung up loosely around the stems. Little mountain orchids are also plentiful in the undergrowth beneath the bamboo, including purplish spidery clusters of *Neottia nipponica,* and a green butterfly orchid (*Platanthera ophrydioides*) that is unfamiliar to us. Rather more robust than either of these is the keyflower orchid (*Dactylorhiza aristata* forma *punctata*), which has spotted leaves and pointy mauve flowers and looks like the common spotted orchids I'm familiar with at home in Northern Europe. The trees rise to shoulder height and the blue hilltops flicker out of view. Here in the dappled shade, a prize find is twinflower (*Linnaea borealis*), a relative of honeysuckle found in subarctic and cool temperate forests. It has tiny, paired, downward pointing bells the colour of apple blossom held atop wiry red stems. This circumpolar alpine is far-flung around our planet – it also grows across North America, from Alaska to California. I tell our hosts that it grows in the UK too, but is very rare and confined to native pine woods in Scotland; they nod silently. Just before we reach the end of the track we notice something peculiar: a tree being engulfed by a black tide of caterpillars creeping vertically up the trunk. Each caterpillar is as thick as my finger, and they've aligned themselves to form an amorphous great mass. I wonder what type of moth they'll become. We reach the end of the track, ready to re-enter civilisation, and leave the countless plants and animals to go about their business in full possession of their mountain once more.

> *As we toiled up the ridge of slippery granite, the three great peaks of Shirane-san rose on our left (west) beyond the deep dark cleft of the glen of the Norokawa, whose subdued roar only served to enhance the silence and the solitude. Pushing on through coarse bamboo grass* (Sasa) *and stunted firs we gained the top of the Tsuetate-toge (7,200 ft.).*
> Walter Weston, 1918.[33]

OUR RYOKAN IN the forest is a rustic affair. By rustic I mean it has no furniture. Not even a chair. We have tatami mats to sit on and a futon each to sleep on, *yukata* (cotton robes) and *geta* (clogs) to wear, but what else would we need? After settling in, we exchange our shoes for slippers at the *genkan* where we're greeted by the two ladies, identically slight of stature, who are looking after us during our stay. They have prepared an elaborate *kaiseki ryori* supper comprising little black plates and bowls of *suimono* (clear soup), shredded *daikon* and various leaves and vegetables, followed by assorted pickles, *takenoko gohan* (rice with bamboo shoots) and a miso broth. We're famished after the trek and our healthy appetite goes down well with our hosts, who bow and chant *thank you please* in unison, giggle to the backs of their hands, and glance at us with curiosity when they think we're not looking.

A purple emperor butterfly (*Sasakia charonda*) in Hokkaidō, Japan.

After supper Ben and I wander along the straight road where crickets purr beneath a pink-washed, painterly sky. The ditch that runs parallel to the road is choked with rough potato vines (*Metaplexis japonica*), which are common around the hill forests of Hokkaidō and smell sickly sweet in the balmy evening air. Their curious dirty-pink, starfish-like flowers grow in clusters that jut out in all directions on thick green stalks. Suddenly I catch a flash of royal blue that could be mistaken for a sweet wrapper (if Japanese roadsides had a litter problem, which they do not). I crouch to get a closer look and, to my delight, discover it is a purple emperor butterfly (*Sasakia charonda*) sitting on the grass; I crouch to admire him, and with little encouragement he creeps up onto my fingertip with wings folded to reveal buff-coloured marbling on the undersides. We saunter along, batting away clouds of gnats, and pass an old woman with crab-like hands and a brown, weathered face. We nod politely and she ushers us over vigorously. She's oddly insistent that we should each take a cherry. '*EAT* it!' she bleats, thrusting the fruits at us vigorously with a weather-beaten arm. We each accept one meekly and retreat to the *ryokan* where I discover a lifetime's mosquito bites on my legs.

> *During our journey down, and in this rainy season, we were molested by gnats* (Culex irritans) *which particularly disturbed us in the night, and sometimes prevented us from sleeping. We were therefore under the necessity of purchasing a kind of porous green stuff, for curtains, such as is used every where in the part of the world, for a defence against these blood-sucking insects.*
> C. P. Thunberg, 1775.[34]

EVER BEEN IN a swamp? I don't mean teetered around the edge of one, or sauntered along a boardwalk in it with a pair of binoculars. No, I mean marched right into it. Bathed in it. Been swallowed in it, bitten by it. It isn't for the fainthearted, I can promise you that.

We challenge the four-by-four to the very limits of its spec, jolting, stalling and revving through a web of Japanese knotweed, and finally conking out in the middle of the black forest. I hope we can escape the place when we need to. Sakaue-san and Kimura-san have mustered up a French gap year student from somewhere who will be joining us for the day. Perhaps she's been promised a taste of 'conservation in action' by joining our expedition – something for the CV perhaps. I wonder if she knows that it will take place in a mosquito-ridden swamp. Anyway she has an impressive grasp of Japanese and is keen to help us with our survey work. The five of us assemble around the car boot, basting ourselves in greasy insect repellent and helping one another affix our nets and helmets. Despite all the body armour, within minutes a horsefly has bitten my upper arm, where a hot pink mound is blooming angrily. Sakaue-san unearths a strongly aromatic green ointment in a little jar from his bag, which calms it down. In a cloud of camphor and repellent we push into the morass.

Let the assault course begin.

Unseen insects scream at us to get out. Whispering and whining, waxing and waning, their cries rise like a crescendo of tuneless violins. But it's the ones you can't hear that you need to worry about. Horseflies the size of cockroaches suddenly pop up along my arms and, one by one, I smack them off with grim satisfaction. Giant butterbur (*Petasites japonicus*) with leaves like upturned umbrellas rise up to our shoulders. Gnats hang in mid-air like plankton. We sink deeper and deeper into oblivion, the black liquid earth sucking at our feet. We're being swallowed alive. *ENOUGH!* we shout, after 30 minutes, by which time it's all become unbearable. We retreat with armfuls of wand-like branches and drooping leaves that we throw into the car and scramble in, along with six captive horseflies that end their lives swotted along the dashboard.

Back at the ranch we dump the greenery on the table and set to work identifying the wilting heap. The room fills with the fresh smell of earth and wet bracken. Once the great mound has been codified on paper, we explore the local forest. It has the feeling of an open, scrubby woodland and overlooks a plain that is sliced from the sky by a ribbon of blue mountains. In the scribbled shade – the sort of place a foxglove might grow – we spot handsome spikes of giant lily (*Cardiocrinum cordatum*) rising above our waists. They're like triple XL versions of a more conventional ornamental lily, with sturdy green stems, each bearing half a dozen illustrious, lime-tinged ivory trumpets pointing in all directions. I've never seen a giant lily in the wild before. It feels like eating ice cream for the first time, or something like that.

WE HEAD NORTH of the Sorachi river to the forests of Nishitappu, where we plan to collect seeds of a plant not yet banked or grown in gardens anywhere. We leave a long, empty road to join a bumpy, suspension-defying track that plunges deep into the heart of the green unknown. Unseen raptors caw above our heads. Everything looks the same for miles, but Kimura-san knows exactly where we are. *Koko*, he says – 'here' – so we park the car and hop onto the track. The air is laced with fern, fungus and little flying insects that catch the light like lint. Fresh foliage has formed a soft green mantle over everything like snow and there isn't a patch of bare

Giant butterbur (*Petasites japonicus*)
in the forests of the Daisetsuzan,
Hokkaidō, Japan.

earth to be seen. We gather up our things and dive into an ocean of giant butterbur parasols that wobble out of our way. We're told that, according to local folklore, fairy-like creatures use the leaves as umbrellas. I can appreciate why: the shadow-webbed canopy at our shoulders is magical. Slender, pale trees shoot up to the heavens, their trunks as straight as spears. We can't see the forest floor, but it feels soft and spongey beneath our feet. At the butterburs' shore we reach the very depths of the forest where the floor is bare and black, like an abyssal plain. Only sparse cylinders of hazy light filter down from the treetops. It's damp and claustrophobic. This is the place, we're told – this is where we need to look.

Almost no one sees the blossoming chestnut under the eaves.
Matsuo Bashō, 1664–1694.[35]

It is astonishing how Kimura-san knows the exact spot where it grows in this vast hinterland, mile upon identical mile of it. But he does. He freezes and points down to the floor. Here, where vines loop down from the gloom, nestles our plant: *Phacellanthus tubiflorus*. There was little chance of us finding it without Kimura-san, for each domed flower cluster is no larger than a plum and they all seem to have gravitated to this spot in the forest. We crouch to peer at the curious white fascicles rupturing from the foot of the vines. A dozen are in bloom, each with scaly stems and toothy flowers that grimace up at us through the leafy detritus. I smile back. They're probably not everyone's idea of beautiful, but I'd take these over a blowsy giant lily any day! We select the most mature specimen we can find, hoping it contains viable fruits for seed-banking, and sever it from its fleshy stalk. Further along the track, we conduct another vegetation survey for completeness. There being no field station anywhere nearby, we resort to transforming the car into a makeshift pop-up herbarium, which works well enough. Like leafcutter ants we file into the forest, raiding the vegetation for snippets and then plastering them to the car bonnet. The omnifarious leaf shapes – jagged circles, triangles, diamonds – together look like a strange, unsolvable jigsaw puzzle. We clock at least 50 species in under an hour.

Our next station is on the edge of the Daisetsuzan region, but feels no less wild. We cut a path through ranks of bamboo and bore into a dark and mysterious swampy forest. Giant ferns spray out from the liquid floor among fallen black branches the shape of antlers. We soon clock yet another butterfly orchid (I've lost count), this one called *Platanthera sachalinensis*; nearby our hosts point out a large Japanese angelica (*Aralia elata*), an unlikely relative of ivy with great wands of leafy, light-catching diamonds.

Field work in the forests of the Daisetsuzan, Hokkaidō, Japan.

Phacellanthus tubiflorus in the
Daisetsuzan, Hokkaidō, Japan.

Pole-like arisaemas jut out from the grasses; everything seems luxuriant in this black, syrupy place. Kimura-san finds a 'sea gull vine' (カモメヅル) (*Vincetoxicum sublanceolatum*), which has little greenish flowers that look more like starfish than gulls. Either way they are beautiful.

There's something else we want to see before we leave Daisetsuzan for the city of Asahikawa. A ghost. More precisely a ghost orchid (*Epipogium aphyllum*). Although famously sporadic, the plant is quite widespread globally. In Britain it's notoriously rare and is indeed intermittently declared extinct before it reinstates itself with desultory appearances; often in beech and bluebell woods set against quaint English village greens and the sounds of cricket bats, clapping and cheering, that kind of thing. The woods around Henley-on-Thames, where I used to live, are a former stronghold. Every year I'd go wayfaring around the woods looking for it, but I never saw one there; few people have. Here in Japan, it haunts an altogether different sort of place, apparently . . . because ominously our hosts are busy putting on dark green rubbery gloves and waders that look fit for a nuclear reactor.

So the four of us, dressed variously for radioactivity or a woodland saunter, march off into the green. Kimura-san leads the charge and, not for the first time this week, has to machete a way into a pathless place guarded by towering weeds. We dissect the broad curtain of rangy stalks and grasses that quiver in a line above our heads. After tunnelling through these, the forest turns out to be more penetrable than I'd feared. The open floor is strewn with large rocky boulders and bamboo detritus, as if a river has run through it. It's rough terrain, but not the worst we've met with this week. Every so often we haul ourselves onto a boulder, pausing every now and again to help one another up. I spot a newly emerged white butterfly with a swollen, red-spattered abdomen, clinging to a branch. Diagonal shards of dusty light pierce the canopy and torch the curly edges of dead leaves at our feet. I can see now how the plant might grow here, for the place

isn't so very different to its British haunts after all. The plant was christened the 'ghost orchid' by botanists Rex Graham and David McClintock who, some 5,500 miles (8,850 km) away, were struck by how it mirrored the dead beech leaves in the changing light.[36] *Torakichiran* is what Kimura-san calls it; I wonder if it means 'ghost' here too.

After about 20 minutes we reach a rocky stream choked with rafts of bamboo detritus. Here's where I saw it, Kimura-san declares, pointing to the opposite bank. We have every reason to believe him – he's not been wrong yet. But the problem with trying to find a ghost is that they're illusory – they hide from you, taunt you – because just when you think you've spotted one you blink, and it vanishes in a trick of the light. It's no less elusive in Japan than in Henley, it seems. We comb the area forensically, kicking up the leaves, parting the branches and staring down at nothing. After half an hour we admit defeat and head back to the *ryokan*. I'm disappointed, but oddly reassured: if it could be found easily, it wouldn't be a ghost orchid, would it?

The hills were not very far from the highroad, and scattered with numerous pools. It was the season of a certain species of iris called katsumi. So I went to look for it. I went from pool to pool, asking every soul I met on the way where I could possibly find it, but strangely enough, no one had ever heard of it, and the sun went down before I caught even a glimpse of it.
Matsuo Bashō, 1644–94.[37]

Searching for ghost orchids (*Epipogium aphyllum*), Hokkaidō, Japan.

HONSHŪ 本州

We have arrived just in time, say our hosts excitedly, as we all bow in the hot morning sun. Nakata-san, Director of Toyama Botanic Gardens, is a small, neatly dressed gentleman in his fifties with a wise face; botanist Godo-san wears thick glasses and an inquisitive expression. 'Come, come!' they say. They lead us into the Gardens' 'sunlight hall' where, paradoxically, our eyes adjust from the glare. What awaits us there is astonishing: the hall is a living forest, packed with long rows of trestle tables draped with sky-blue cloth that form the stage for hundreds of bonsai trees. Their gnarled, stunted trunks are about as thick as an ankle, twisting their way out of small black trays with gold Japanese characters engraved along the front. The trees are covered with puffy clouds of pink, white and peach blossom, forming little pastel-coloured explosions around the room. '*Satsuki* azaleas!' exclaim our hosts, beaming. Some of them are centuries old and have been handed down for generations, they explain. With sadness, they tell us the next generation won't know how to care for them; 'No interest. No interest,' they repeat disconsolately. Ruminating on this, we're ushered to the corner of the room to see a 'tortoise', which turns out to be a potted *Dioscorea*. It has a hexagonal, wooden hump at the base of its stem that looks remarkably like its Japanese namesake.

In the afternoon, we change into our suits. I feel like I should be changing into my pyjamas after the international flight, but sleep is out of the question because we have something important to attend. This afternoon we have a meeting scheduled with Ishii-san, Governor of the Toyama Prefecture. The meeting – a small, private affair we're told – is in honour of the collaboration between the botanic gardens here in Toyama, and our own. We pull up by a large, cream-coloured municipal building with files of black windows overlooking balls of topiary. Our translator – a man in his twenties – whisks us away to a sizeable waiting room where we sink into low, white leather sofas and make polite conversation over cups of iced green tea. After 15 minutes we're taken to an identical, if somewhat larger, waiting room where the process is repeated. And then to a third. By the time we reach the fourth, the sofas have become the size of super king beds and we've drunk dangerously large volumes of green tea. Now we're ready to meet the Governor, announces a small, smiling, smartly dressed lady. His office is positively palatial. We file in, bow and take our seats. Our small, private meeting with the Governor has an audience of 30 journalists, all dressed smartly in black and white suits, who click and flash from a metre away while the speeches are delivered and translated. After a brief TV interview and 16 more photographs staged outside the office, we're spirited back to our hotel, desperate for the loo, and with the negatives of a thousand camera flashes imprinted on our eyes.

THIS MORNING OUR hosts have planned a tour of botanic gardens. We're joined by the Director of the University of Oxford Japan Office, Alison Beale. Alison is a gifted linguist who worked previously in international relations for the British Council, and has spent most of her life in Japan. She translates our discussions, pausing every now and again to query a botanical detail. We sit in Nakata-san's office drinking green tea and chatting happily about the plants depicted in heavy tomes of the country's rich flora that he pulls one by one from his shelf. Before the tour can begin, Charlie must be located. Charlie (whose Japanese name we have never been told) is a tall, dark-skinned man with a wry grin. Curator of the Gardens and a gifted plantsman, he rules a kingdom of glass prisms containing a thousand pots, tanks and baskets of botanical treasure. He is one of life's eccentrics. Duly spotted in a dark corner of the sunlight hall with his *shamisen* – a traditional three-stringed instrument with a loud twang – he insists on singing us a quick Japanese folksong before the tour is set in motion. When he's finished, he takes a bow, and we all pile into an electric buggy, curiously said to accommodate either 14 or eight people (depending on their size, I presume). There is plenty of room in the buggy, (regardless of size), but Charlie opts to take his bicycle – a shade too small for him – and slowly pedals alongside us.

Soon after setting off we see the first plant of interest: *Monotropastrum humile*. It's a leafless, ghostly white plant, each stem supporting a nodding flower that looks strangely like a pony's head. We see herds of them pushing their way out of leaf litter by the path. It's parasitic but, unlike broomrapes and *Hydnora*, both of which feed from the roots of other plants, this species obtains its nutrients from a fungus. We discuss how scientists in Japan have unexpectedly just discovered that its seeds are distributed by cockroaches. It has equally unusual relatives that lurk in forests elsewhere in the world, such as sugarstick (*Allotropa virgata*) in the Pacific Northwest. Its spikes of flowers rise from the leaf litter on stems that look extraordinarily like red and white striped candy-sticks.

Next, the buggy glides to a large lake fringed with native Japanese water iris (*Iris ensata*). Their silky, ametrine flowers unfurl from long, slender stems that look too flimsy to support them. Curious to learn more about the provenance of the lake, and not wishing to bother our hosts with too many questions, I try to translate the Japanese text on the tourist leaflet using the translator app on my phone Ben has shown me. I scan through the things I can 'see, learn and enjoy' here, of which there are many. The translations are hit and miss, I'm learning. For example, among the various 'items that may cause inconvenience to other visitors', I discover 'shooting them, without permission'.

TATEYAMA, AT 3,015 m (9,891 ft) belongs to a range dubbed the Japanese Alps, and is one of the 'Three Holy Mountains' of Japan. Today we're participating in an excursion to explore the alpine flora of the mountain's lower slopes beneath the Murodo Plateau, at 2,450 m (8,038 ft). We take a part of the Alpine Route which doesn't permit private cars. Besides designated buses, ours are the only vehicles. We snake up steep, softly forested hills and park in the dappled, lime-green shade. The air smells deliciously cool and fresh, like woods after the rain. We wander along the roadside peering up into the canopy, examining a leaf or crouching to inspect the roadside undergrowth. A multitude of mountain roadside flowers grow beneath the towering cedar trees (*Cryptomeria japonica*). We assemble around a large, venerable specimen of cedar to have our photograph taken. HAI, HAI, HAI! (はい) we all snap in chorus. Here among the cedars our hosts point out a section of forest that has been cordoned off, where a bear attack happened last week. Bear attacks are a part of life here, despite the animals' shyness of people. In some years over 100 people are attacked across Japan, and trekkers on Tateyama are advised to be cautious. 'Is there anything we can do to ward them off?' I enquire. One of our party forages the supplies in the back of the vehicle and I anticipate him whipping out a Samurai sword or similar. A tiny bell emerges. Perhaps they detect our misgivings because they promptly explain that bear bells are the best way to warn the animals of a trekker's presence and avoid an attack.

> *Cedars (*Cupressus japonica*) grew in great plenty hereabouts, as well as in most of the other provinces; but nowhere, perhaps, can they be found finer, or in greater numbers . . . We now left this beautiful spot, and proceeded on our journey down the mountain, during which time I did not neglect diligently to search for and collect the flowers and seeds of the plants and bushes that grew by the road-side.*
> C. P. Thunberg, 1775.[38]

As we jangle along the road, the air becomes thick with the scent of wild magnolia. The appropriately-named 'Japanese big-leaf magnolia' (*Magnolia obovata*) has flopping great leaves the size of dinner plates, which are apparently used for wrapping food. Their white, bowl-shaped flowers are equally impressive. There are exceptional plants wherever we look: wild acers (*Acer ukurunduense*) like delicate sycamores fan out above our heads, their branches quivering like feathers; beneath these are pink dollops of weigela blossom (*Weigela hortensis*) like strawberry ice cream and, in the lowermost tier, a patch of Asian royal ferns (*Osmunda japonica*) pushes out clenched fistfuls of cinnamon-coloured, spore-producing fronds from the thicket. A few metres from this botanical bonanza, an unexpected find lies at our feet: a rare silver orchid (*Cephalanthera erecta*). A flock of white flowers circles its stem like tiny swans. I'm more familiar with seeing *Cephalanthera* orchids in light chalk woodland and downs than in dense forest undergrowth on clayey soil.

There's more. We park the car again at 1,400 m (4,593 ft). The verge is saturated with meltwater that has pooled around soggy, straw-coloured tumps, out of which young horsetails are poking like asparagus. We squelch along the embankment. Smiling, our hosts point towards a patch of one of the most beautiful plants I've seen in all Japan. You wouldn't think so from its name: white skunk cabbage (*Lysichiton camtschatcensis*). Yes, skunk cabbage. It's named after its distant cousin, the western skunk cabbage (*L. americanus*), which has malodorous flowers more deserving of their title. Here, the Japanese call it *mizubashō* meaning 'water banana', which is moderately better I suppose. 'A rose by any other name . . .'. There are about a dozen in full bloom, each with a spathe like satin, enclosing a yellow, club-shaped spadix; each one perfect and unblemished against the sludge. We crouch to admire them and take photos, sinking into position.

Even the skunk cabbages can't prepare us for what we see next. A steep, narrow track transports us into a mountain scene that belongs to a dynasty painting. It's breathtaking: slender, leafless anise magnolias (*Magnolia salicifolia*) arch above our heads scattered with large flowers like silky white sashes tied to the branches, interlaced with leafy fans of white-flowered hydrangea. Both are splashed with cerise azaleas (*Rhododendron albrechtii*) – together they look like drifts of strawberry and vanilla ice. All this frames a view without compare. Black rocky peaks criss-crossed with snow rise out of a vast forested wall from which the Shōmyō Falls tumble down into the mist. *Sugoi!* (すごい) exclaim our hosts – 'incredible'. We spend a glorious half hour in this magical place before we weave back down the steep little track. Along the way we spot pink snowbells (*Soldanella* sp.) pushing their way assertively out of the earth. It's difficult to make sense of how much natural beauty we've seen in a short space of time. Quietly overwhelmed, we continue our ascent to the Murodo Plateau.

> *Over the edge of a deep cliff in the centre of a mighty tree-grown amphitheatre, a far-off cascade, at the head of the Shomyo-gawa, falls in three great leaps a thousand feet.*
> Walter Weston, Mount Tateyama, 1918.[39]

Anise magnolias (*Magnolia salicifolia*),
Mount Tateyama, Honshū, Japan.

1,500 m (4,921 ft). The landscape changes dramatically as we
enter the montane zone, and the vegetation thins. Great triangular
peaks jut out on each side of the road; they look draped in black
and white camouflage. Then we enter a canyon unlike any other.
Yuki no Otani, 'The Great Valley of Snow', tunnels through some
of the world's heaviest snowfall. We disappear between two white
walls and it feels as if we're burrowing into a glacier. The snow
can reach 20 m (66 ft) and the walls are still 16 m (52 ft) high
in early summer when the road is ploughed – taller than a five-
storey building. At the other end of the blue and white chasm
the snow recedes to reveal our destination: the volcanic Murodo
Plateau at 2,410 m (7,907 ft). This is the base camp for those
planning to summit the 'roof of Japan' and it's a hive of activity.
Throngs of Japanese tourists in mountain gear flash smiles and
peace signs to phone cameras besides every rock, building and
vehicle; one group is even staging a photo by the loos. We put on
our sunglasses to protect our eyes from the intense glare, and set
off on foot to the *onsen* where we are to have lunch. The rope-
lined path dissects lush alpine meadows that clothe the plateau.
After a short trek, on our right we pass the Mikurigaike Pond
– a serene pool filling a crater to form a perfect mirror for the
surrounding peaks. Jagged white shapes drift silently across its
still blue surface. As we continue along the track, to our left, the
land falls from our feet. We've reached Jigokudani 地獄谷, 'Hell
Valley'. Plumes of steam and volcanic gas pour out of the abyss as
we stare down into the underworld.

After lunch at the *onsen* we clamber up into the alpine
meadows above the valley against a cold blue sky studded with
clouds that mirror the snow drifts. The dark green brush is short
and thick; it barely rises above my ankle. Summers are short
here, and the conditions are challenging for plants to survive
in. Where the snow has thawed we observe patches of white
anemones (*Anemone narcissiflora*) unfolding. We congregate to
admire these before starting our descent back to the Murodo

base camp – it's getting cold, and the party is weary. We pick our way over the rocks and chat about all the beautiful plants we've seen.

Then something extraordinary happens. Out of the depths of Hell Valley soars a great 'thunder bird' 雷鳥 (a rock ptarmigan), which flaps to a halt at our feet. He snatches at an old stump with hairy-white, griffon-like feet and twists his head to gaze out over the mountains, seemingly oblivious to our presence. His pinion of brown and white smudges is the embodiment of camouflage – had he not moved, he would be invisible against the thawing snow. In a panic, he vanishes, and we're left wondering if he was ever even there.

> *But the extra-ordinary beauty of the romantic glen*
> *surpassed anything I had previously met with in Alpine*
> *Japan . . . it was a constant and kaleidoscopic succession*
> *of charming scenes. Its chief feature was the splendid*
> *fuchi, deep pools of emerald green, now and then opening*
> *out from some narrow, rocky defile, the home of the*
> *iwana, each one more beautiful than the last.*
> Walter Weston, 1918.[40]

A BABY GREEN dragon gawps at me from the edge of the car park. Our convoy of vehicles crunches to a halt on the gravel and car doors slide and slam; I wander over to take a look. *Pinellia* is his scientific name, he being an aroid, and I say 'baby' for his chalky-green hooded spathe is no larger than my thumb. Out of

A green dragon (*Pinellia* sp.),
Miyagawachohora, Hokkaidō, Japan.

Wisteria vines in the temperate
rainforest at Miyagawachohora,
Honshū, Japan.

his mouth pokes a long, shiny, mouse-tail tongue – his spadix. I peer into the velvety chocolate interior and smile. I've never seen a green dragon in the wild before and to find one from the get-go here in the car park bodes well for the day.

A remote tract of temperate rainforest in Miyagawachohora is the target of our botanical survey work today. The heatwave we were warned of has ripened to a very humid 35°C (95°F). Our hosts are all swaddled in towels – 'so hot, ever so hot!' they lament repeatedly. We trundle down the wooded path, fanning ourselves with leaves, clipboards and – mysteriously – a plastic chopping board, and then huddle around a pile of rucksacks to survey the wall of vegetation rising in front of us. A specimen of every vascular plant needs to be collected from this vast, pathless wilderness. It won't be easy, we surmise, but there are seven of us and we have all day. After a short colloquy on how best to divvy up the mammoth task ahead, we fan out with eyes fixed to the floor like a forensic team. The forest hisses. The noise and sultry conditions make the rainforest feel more tropical than temperate and the place is crawling with bugs. I notice a striking cream and black butterfly fixed motionless to a leaf. Without time to admire it, I press on, forcing my way through a curtain of wisteria vines. I'm familiar with these plants in a quaint English country garden setting, but here in the indigenous forest they are something different: fat grey pythons wrapped around the trees, slithering in and out of the branches. I can't see their flowers; presumably these grow far above our heads in the canopy. I come face to face with a large yellow and black spider as I wipe the sweat from my brow and head deeper into the gloom.

Every now and again we cross paths (for we must forge our own) and startle one another, then compare our leafy bouquets. Near the scribbled edge of the woods I encounter a small butterfly orchid (*Platanthera minor*). In Japanese, its name means 'dragonfly grass', which sounds beautiful. Its curious white flowers look more like little doves than they do butterflies

A specimen of *Metanarthecium luteoviride* collected in the
temperate rainforest at Miyagawachohora, Honshū, Japan.

or dragonflies. After about an hour we amass our specimens in a great heap, which is stuffed into a polythene bag and carted off for identification. Hot, sweaty and criss-crossed with scratches, we wander back to the lodge. On the way I ask our hosts for the name of a plant I don't recognise growing beside the track. It has a rosette of leaves that might belong to an orchid, and a spike of small, spidery green flowers. They promptly identify it as *Metanarthecium luteoviride*, an unusual species native to the damp forests of Japan, Korea and the Kuril Islands. Nearby we see banks of wild hostas and epimediums, as if someone had planted a woodland garden here in the Japanese wilderness.

Mercifully, the forest lodge is air conditioned. We take off our shoes and deposit our cargo on a long trestle table. As the party chew on their triangles of rice and sip iced green tea I wander around the lodge, which is some sort of natural history museum. Stuffed animals snarl from glass boxes and a fat grey frog stares at nothing from the bottom of his undersized tank. Cool and sated, we set to work identifying the samples. We upend the polythene bag by two corners and a leafy stream pours onto the table, along with various critters that either hop out or zoom off. We find a small caterpillar in the catch that looks impossibly like a twig. Naturally, we have duplicated many plants seven times; each of these is baptised *onaji* (同じ) meaning 'same' and swept into a separate pile. The remainder we hold up to scrutinise one by one, carefully examining leaf shape, hairiness and various other attributes. Ohara-san, one of our hosts, is a living field guide to the flora of Japan. Patiently he nods and smiles generously when any of us makes a correct diagnosis, or looks quizzically at the plant and politely offers an alternative suggestion when we don't. He's right every time; there isn't a single specimen to which he cannot assign a name. The verified plants are stacked between sheets of newspaper and bundled up with cardboard and string, ready to be dispatched to the Oxford Herbaria. By the time we've finished, it looks as though somebody has detonated some kind of plant bomb – vegetable matter is even poking out of our hair. We do the best we can to tidy up, mumbling our apologies to the keepers of the lodge who nod solemnly as we cross the threshold from this cool sanctuary back into the hissing wilderness.

In the afternoon we're taken to a nearby marsh to examine the aquatic flora. We pass vertical orange banks splashed with salad-green ferns (*Blechnum nipponicum*) and spot a stout, black and green marbled stalk of *Arisaema serratum* var. *serratum* rising out of a bamboo-clad ditch. At the top of its stem is a fist of developing green fruits, like foam. We emerge from the closed-canopy forest at a lush, open, grassy glade by a lake lined with softstem bulrushes (*Schoenoplectus tabernaemontani*). From the water's edge we're shown the buttercup-like flowers of a rare Japanese waterlily (*Nuphar saikokuensis*) that was described

Processing specimens collected from the temperate rainforest at Miyagawachohora, Honshū, Japan.

An *Epimedium* specimen collected in the temperate
rainforest at Miyagawachohora, Honshū, Japan.

A *Senecio* specimen collected in the temperate
rainforest at Miyagawachohora, Honshū, Japan.

as a new species in 2015. Large silky leaves of sacred lotus (*Nelumbo nucifera*) flop out of hidden corners around the lake. Best of all, on the way back to the car we discover a patch of pink lady's tresses orchids (*Spiranthes sinensis*) peeking out of the grass. They look just like the white-flowered forms I'm familiar with in northern Europe, only more colourful. The dainty rose-coloured flowers spiral 360 degrees around their vertical stems like miniature stairways.

On the way back to Toyama we make a diversion to see a rare and special plant. It was mentioned over lunch and, seeing how excited I was by it, our hosts have kindly taken me to see it. We park at a layby and scramble up the roadside verge. Each of us forages for the plant, parting reeds and peering down at the boggy ground. A seep runs down the bank and the earth is green with algae: perfect conditions for the plant we're pursuing. Sure enough, a few metres from the road beneath us, half a dozen wild sundews (*Drosera tokaiensis*) are proclaimed. Their small, ground-hugging rosettes of paddle-shaped leaves are reflexed over the ground and glisten with dewy, ruby-coloured tentacles. This roadside is the northernmost locality at which the plant occurs in Japan, I'm told. What better way could there be to end a day of botanical surveying than to see a carnivorous plant in its natural habitat?

> *The* Nymphea nelumbo, *in several places grew in the water, and was considered, on account of its beautiful appearance, as a sacred plant, and pleasing to the gods. The images of idols were often seen sitting on its large leaves.*
> C. P. Thunberg, 1775.[41]

UMEBOSHI, SAYS OHARA-SAN, smiling. 'For headache.' He hands me a small lunchbox of pickled plums in brine. He knows that a party of us went out to celebrate a week of successful survey work last night. He might or might not know that I went toe-to-toe with Charlie at karaoke and we all took shots into the small hours. Either way he's guessed that I'm not feeling my best this morning. He drives me to the botanical gardens where we meet Godo-san in the car park and the three of us head off to the banks of the Jinzū river. From the bridge we see fishermen with straw hats, up to their shoulders in river water, casting long rods. *Ayu*, states Godo-san, only repeating the word and nodding affirmatively when I seek further explanation. A river fish served up as a delicacy in all the restaurants around here, I learn later. The road meanders down to the river and we park on the grassy bank under a low, cotton wool sky. I chew reluctantly on my *umeboshi*, which so far have done nothing to ease the pain.

My hosts have kindly given up their Saturday morning to take me here to look for a rare broomrape (*Orobanche coerulescens*). The three of us plod along the riverbank's gravelly path, pondering over plants we encounter along the way. My clammy tee-shirt sticks to my torso and sweat trickles down my back. It's uncomfortably hot down on the banks of the Jinzū today, with or without a hangover. Godo-san has wrapped a blue towel around his head like a turban. Soon after we set off we spot robust clumps of orange-flowered daylilies, like the popular form grown in gardens, growing wild on the plain. Among its pleated leaves lurks a sizeable spider with ominous, wasp-like markings. My hosts assure me that 'Jorō', as they refer to her, is harmless; nevertheless, I'm in no mood for a venomous bite today so I give her a wide berth.

After about half an hour, Ohara-san declares broomrape-hunting season open, and the three of us fan out from the path, intent on the ground. To my excitement, in minutes we stumble across the plants pushing out among the grass and wormwood.

Orobanche coerulescens
on the banks of the Jinzū
river, Honshū, Japan.

A Jorō spider on the banks of the River Jinzū, Honshū, Japan.

About a foot high, they have thick, fawn-coloured stems clothed in white hair, topped with clusters of tubular flowers the colour of violets. We crouch to get a better look at our reward, peering at them sideways on our hands and knees. 'Hooohhhhh?' chant my hosts quietly in unison – an expression tantamount to 'gosh, that's something', but with the intonation of a question. We select a suitable voucher specimen and then look for one in fruit, for Godo-san fancies that he might try growing the plant in a petri dish. I take photos, sketch the plants and generally just enjoy time in their company under the strong summer sun, with just a breath from the Sea of Japan a few miles away. My headache has disappeared; perhaps it's our discovery, or maybe the *umeboshi* are kicking in at last.

A hangover: but while the cherries bloom, what of it?
Matsuo Bashō, 1670–9.[42]

GODO-SAN AND OHARA-SAN invite us to join Toyama's Hashi Matsuri festival. The festival, they tell us, dates to 1869 and celebrates the completion of the Tozai Bridge across the Shiraiwa river. Upon arrival, we join a long chain of cars combing the road for parking spaces that don't exist. Eventually we abandon ours in a side street and set off along the sultry banks of the Shiraiwa with the crowd. Ben and I look out of place here, not just because we're the only Westerners, but because most attendees are dressed in striking traditional Japanese garments. Perhaps that's why they're staring at us. Ohara-san is wearing an elegant, vanilla-coloured *yukata* with a black sash. Young men are clad in sleek black *yukatas* while the girls have colourful cotton ones, tied with complicated sashes at the back, and flowers in their hair. I feel as though I've stepped into an eighteenth-century Ukiyo-e painting. Along the banks of the river, long lines of red and white paper lanterns and ribbons flicker and twist on the breeze, glowing against the low grey pall. The air smells of summer evening grass.

Eventually the moving human kaleidoscope clicks into position and we find ourselves on the middle of the bridge. Everyone has turned to look upriver where a small square stage has been set afloat. Like everything else (including a half-sunk fishing boat), it's festooned with glowing paper lanterns. The stage looks decidedly aslant, but nobody seems bothered by this. Now, we're told with a smile, 'Hinagashi' will begin. What we're about to see is truly special: a thousand box-like paper lanterns, decorated with 'wishes', are lit upon the platform and released floating into the Shiraiwa and out into the bay. It's a hauntingly beautiful spectacle that mesmerises us all. As night falls, we drink beer and eat *yakitori* sticks in the sticky heat by great yellow *aka-chochin* lanterns. At the end of the ceremony, Godo-san takes us to a nearby shrine. Upon instruction, we throw coins into the offering box, ring a bell hung on a thick cord, clap our hands and then bow deeply. As we wander back to the car, the last of the lanterns are still twinkling their way along the black water like a river of stars.

A special meal has been planned here in the city of Toyama. It's one of many that our generous hosts have arranged in celebration of the important work our gardens are doing together. We take off our shoes and file up the stairway of the Kaiseki-ryōri restaurant into a room screened by paper. We sit on the floor around a large rectangular black table. A procession of glossy seafood commences with bowls of firefly squid. These are a speciality of the Toyama Bay that, in life, light up blue like glow sticks as they dart about the sea. They droop from our chopsticks as they are dangled steadily into our mouths. Saké is poured for one another (one mustn't pour one's own) from sizeable bottles (one cannot have too much) while our hosts talk excitedly about the prized delicacy of the meal. Genge is a rare and special fish that has been dredged from the very depths of the ocean in honour of the occasion, we're told solemnly. The brown, frowning creature is duly wheeled out and I declare to our hosts that I've never encountered such an unusual fish (of course I haven't, it's found only in Toyama). We all agree it has a most agreeable flavour. Next up is a tea party of *chawanmushi* – a form of custard, I'm told, served in little ornate cups. Certainly it looks a lot like custard. Therefore it's rather a surprise when the sweet desert I'm anticipating tastes of salt and smoked mackerel; it's even more of one when I discover an eel lurking at the bottom. I force a polite smile, one that remains frozen on my face as I stare down at the little serpent.

The saké is positively flowing. Especially for those seated near Charlie, who's wearing a scarlet Hawaiian shirt and seems larger than life. 'Have a *driiink*,' he says, looking at my half-full cup that he'd topped up not five minutes ago; 'have *anoooooother* one,' he slurs, sloshing it all over the table. He picks up his *shamisen* and

Toyama's Hashi Matsuri festival at
the Shiraiwa river, Honshū, Japan.

sings us his 'song of the sea'. The rest of the party carry on with their conversations, paying him not the least bit of attention.

We've all brought gifts to exchange – a planned surprise for which the appropriate moment arrives shortly after the third course. The ill-concealed stack of paper bags is brought forth from the back of the room and we carefully apportion our own seven gifts in order of rank, starting with the Director's. Unfortunate. As we offer him the gift, our faux pas is met with a sea of batting hands and cries of 'No, no, no!', along with 20 head jerks to the left, where the Governor of the Prefecture – a late addition to the party and an honoured guest – is sitting. This rather throws us. Anxiously, the gift sequence is reassigned and, much to our relief, the seventh empty-handed player of this complicated game of pass the parcel loses gracefully, so all is well. Charlie announces that he cares to make a speech, and stands up. The din around the table clutters to a halt and we all look up at him expectantly, with chopsticks poised. He clears his throat. I panic briefly, wondering if perhaps we're all expected to make speeches, and draft something in my head about the honour of collaboration and the importance of our work. 'SON. OF. A. *BITCH*,' announces Charlie slowly and deliberately, then sits back down again and knocks back a shot of saké. There's a moment of astonished silence, after which everyone roars with laughter and claps, patting him on the back as he pours himself another drink. 'Charlie learned English from American soldiers in Okinawa,' someone explains to us. I guess we're not expected to make speeches, after all.

Each evening the sake flowed freely in the charming
quarters . . . while nearly all night long . . . a constant
procession moved to and fro . . . singing in the 'bird- like
voice of the Japanese' . . . The exact nature of the bird he
has not specified, but perhaps a night-jar or an owl would
be an appropriate comparison!
Walter Weston, 1918.[43]

SHIKOKU 四国

We take the slow train from Okayama to Kōchi (高知) on the island of Shikoku, shelling edamame beans from their pods and admiring fine views. Pacific-facing, mountainous and dissected by rivers, Shikoku is a watery wilderness. A complicated metal bridge takes us over a Mediterranean-blue expanse interrupted with cone-shaped islands cloaked in black forests that pitch to the sea. On the island, we trundle past paddy fields and pointy, Yosemune roofed-houses among stands of bamboo stems as straight as arrows and large as trees. By the time we approach Kōchi, a chain of low clouds clings to the drippy forest, and then the views become swallowed up by mist.

> The bamboo (Arundo bambos), *which is the only kind of grass that grows to the size of a tree, grew in many places, and differed much in height and thickness. The root of it is made use of here . . . for (Atjar), pickling with vinegar, The thicker stems were used for carrying burthens, and the finer branches as shafts for pencils, and when slit up, for fan-flicks and for many other purposes.*
> C. P. Thunberg, 1775.[44]

AFTER CHECKING INTO the hotel, I stroll along the steamy banks of the sluggish Enokuchi river. Phoenix palms (*Phoenix canariensis*) have been planted along one side, and epiphytic ferns spray out from their moist, shaggy trunks. I disturb a large, ochre-coloured crab, and mumble an apology as he slinks back into the mud. I cross a little bridge that looks out over a jumble of traditional ornate roofs and modern white buildings, with the occasional bonsaied niwaki tree poking out. Everything is unmistakably Japanese.

In the evening we dowse ourselves with aftershave and comb the warm streets of Kōchi for a place to eat. After weeks of seafood I rather hope we can sample something different: meat perhaps. We pass a traditional *izakaya* by the river. The wooden hut is draped with a gaudy assortment of flags, signs, wires, satellite dishes and illuminated lanterns. And, oddly, bicycles. Unable to decipher this, we decide to take our chances here. Inside, a row of serious-looking men are eating at a bar, their eyes fixed to a television screen rigged to the wall amid a tangle of fishing nets, dried starfish and glass tanks containing fidgeting crabs and other assorted sea creatures worthy of an aquarium. The sea slugs especially. 'Ah, here you are,' says a lady in a black apron, as if we were expected. We're ushered to a wooden booth, plastered from floor to ceiling with photographs of a bewildering array of plated marine life. It's a kaleidoscope of seafood.

I may not have opted for seafood, but the *sashimi* is out of this world: a banquet of cubes and triangles of succulent pink and white fish served in leafy receptacles with soy sauce, and deep-fried, salty pieces of goodness knows what. We wash it down with beer and order more – and more. I decline the weighty conical molluscs that still look alive, although Ben seems to relish them. The hostess is a scream: a lively woman in her fifties, she seems fascinated by our very existence (Westerners are scarce here). She studies our hands and faces and says 'Why Shikoku?' 'Well, we're here to . . .' 'Yes, yes, yes,' she cuts us off and then, to the air, snaps 'SAKÉ!', which is promptly brought to us by her obedient daughter. The hostess's English is little better than our Japanese, but the conversation is lubricated by the halted translations her daughter offers, along with the river of saké that is now flowing. '*Oishī*!' (おいしい) –

delicious – we exclaim repeatedly about the cuisine, and each time we say it, another jug is brought to us. She drinks from it too. We're congratulated for our youth, our pale skin, and our fine, large noses. And then, after another animated exchange about the excellent quality of her food, we learn that we're to be considered suitors for marriage here in Kōchi (although suitors for whom it's not entirely clear). 'Not that one,' she says, shooing away a plain girl in her mid-twenties vigorously with one hand. Soon a Polaroid camera is unearthed from somewhere and we become captured in the very fabric of this extraordinary place.

MAKINO BOTANICAL GARDEN is an oasis of lush dells and ponds overlooking the blue mountains of eastern Kōchi that enclose the island's largest alluvial plain. The air smells alive with plants. White wind orchids (*Vanda falcata*) have strapped themselves to the trunks of trees with finger-like roots, not far from the garden's buildings. A miniature rocky cliff is roofed with a thick layer of papery tongue fern leaves (*Pyrrosia lingua*). A stone's throw away, flowering spikes of violet hostas (*Hosta capitata*) and white sprays of astilbe (*Astilbe japonica*) arch over mounds of feather-soft moss by a stream forcing itself through the rocks. Most beautiful of all, a pair of large pink orchids (*Bletilla striata*) nod from the ferncry. Their vibrant, stripy lips are ruffled like crepe paper – they could be made of origami. I stand back and take it in. Like all good gardens, it's unclear where the natural vegetation starts and ends.

A lovely rock-garden, where, half hidden in the azalea
bushes and iris beds, the waters of a mountain cascade fell
into a little pond with ceaseless roar.
Walter Weston, 1918.[45]

We're soon met by Ayako-san, the local botanist, a small lady in her thirties with a shy, earnest demeanour. She leads us through a modern, architectural building to a spacious wooden boardroom. Jugs of iced green tea are brought to us, and we discuss our plans for fieldwork and conservation. We pore over great tomes and maps to plan the best use of the short time we have here. After lunch we're taken down to the herbarium. It's an impressive, wooden-clad vault that smells of old books, preservatives and time. We bow to its keeper, who tells us the herbarium contains some 300,000 specimens, the majority collected from around the Kōchi Prefecture. I ask to see a specimen of the rare and poorly known parasitic plant *Mitrastemon yamamatoi*, which I know occurs in the region. The metal shelves are parted and multicoloured stacks of flimsies splay out from floor to ceiling. The said specimen is located, and we lay it on the table. Parasitic plants make poor pressed specimens and this one is no exception (it doesn't even possess leaves); nevertheless, I'm happy to see such a special plant. We make our way back to the car park where we're joined by another of the garden's botanists, Michiyo-san, who sports white gumboots matching Ayako-san's. We load the four-by-four with bags, water bottles and weighty reference books, and set off to explore the prefecture.

Our destination is Shikoku's great Monobe river (物部川), where we hope to find a population of rare broomrapes (*Orobanche coerulescens*). We drive through a patchwork of fields and villages down to the river bank, where we park. The four of us fan out from the vehicle over the grey shingly banks under a low, cloudy sky criss-crossed with wires and pylons. As we crunch over the gravel I connect the place, improbably perhaps, with the docks in South Wales where I hunted down yellow broomrapes, years ago. However, my radar for broomrapes (on which I pride myself, by the way) doesn't work as it would in Britain, here on unfamiliar terrain. Eyes fixed to the ground, they do not reveal themselves in the places I feel they ought to. I pick my way through tufted grasses and low, scratchy thickets. Childishly, I need to be the first in the group to find one and, eventually, my impatience pays off: a squat yellowish stem, bristling with purple flowers, glows against the dull grassy scrim. I shriek with delight, and everyone scuttles over to look.

We soon find more in the sward. A particularly robust one is plucked for the herbarium while Ben and I busy ourselves with seed collection from crispier fruiting specimens. We take photos and admire the curious plants. Upon closer examination we discover that the population here on the Monobe riverbank is hairless, so we can assign the plants to the Japanese forma *nipponica*; they are distinct from the woollier form that we found along the Jinzu river a week or two ago. Warmth floods my veins like an opiate as we trundle back to the botanical garden's herbarium to stow away our treasure.

THE LONG ROAD to Mount Kamegamori (瓶ヶ森) winds its way
north in parallel with the Niyodo river. On our right, a steep bank
of concrete blocks is furred over with moss. Above it, a wire mesh
is doing a shoddy job of holding back lush vegetation that looks
ready to take back control of the road. We stop to buy *onigiri*
– little pyramids of rice enveloped in seaweed; there'll be little
chance of us finding food later, for we're heading into the depths
of Shikoku's vast mountainous wilderness. After a two-hour drive,
the road narrows and becomes overshadowed as we snake into
the foothills. The foot of the verge is spattered with wet, drooping
ferns, and above it rises a wall of shrubbery and bamboo. Keen to
make sense of this scribble of vegetation, we abandon the car in
the road (for we have the mountain to ourselves), stretch our legs
and explore the dark, misty forest. It smells wet and earthy. To
our delight, a cobra lily, or fancifully-named 'jack-in-the-pulpit'
(*Arisaema tosaense*), jumps out from the forested slope. It looks
ethereal, with a bright lemon and lime stripy spathe that arches
and dribbles over its spadix; it almost glows in the murky forest.
We spend some time photographing and admiring it. It's a star
find, but I'm holding out for something even better today.

> Up a valley of surpassing loveliness, each turn of the
> winding glen more romantic than the last. Here and there
> the track merely rested on struts of timber driven into
> the precipitous crags of dazzling granite, overhanging the
> flashing emerald waters of the Yugawa, with an almost
> sheer drop of 500 ft. to the valley floor.
> Walter Weston, 1918.[46]

We spiral up, out of the foothills, and glimpses of jade green
forested triangles of mountainside flicker through the trees and
swelling clouds. We make another stop to assess the montane
roadside shrubbery. Within minutes of leaving the vehicle, we're
soaking wet, even though there is no rain. Our guide hands us

towels to drape around our necks. We find rhododendrons (*Rhododendron kaempferi*), as drenched as we are, capped with clusters of coral pink bells spotted with crimson, as fine as any I've seen on a cultivated shrub. Minutes after leaving the car, thick white mist pours into the valley and blots out the sun. But stars are shining at our feet. Earth stars. These extraordinary little fungi (*Astraeus hygrometricus*) look curiously like little starfish twinkling among the weeds and rock. We peer at one more closely. It has a frosted orb atop a ring of radiating blue-grey prongs; it seems to belong to another world.

1,000 m (3,281 ft). The air is cool and clammy. Fuchsia-pink weigelias (*Weigelia hortensis*) and white bridal wreath (*Deutzia crenata*) spray out of the roadside shrubbery. We make our way steadily over the south-eastern spur of the mountain and by 1,600 m (5,249 ft) the vegetation is waning. Forest-clad buttresses of rock fall sharply from the road into a misty ravine. These are crested by a carpet of dwarf bamboos (*Sasa tsuboiana* and *S. palmata*) forming a thick, papery thicket around Nikko fir trees (*Abies homolepis*). We pause to sit and chew on *onigiri* in our drenched cagoules at a chilly, wet mountain viewpoint; only by now there is no view because it's been swallowed up by cloud.

After several hours' combing the summit, we head back down the mountain to the richer middle-elevation hill forests to carry out our survey work. We leave the summit road and take a crooked track scarcely fit for a vehicle, so we proceed the rest of the way on foot, towels still draped round our necks. Here in this remote tranche of forest, the air smells sweet, fresh and green, like pines. I'm aware of the rhythm of my breathing, of the forest's oxygen; it's like I'm more alive here. The sky clears and layers of green and blue mountainside concertina out into the distance to our left,

Arisaema tosaense growing in the foothills of Mount Kamegamori, Shikoku, Japan.

framed by sprigs of white hydrangea (*Hydrangea luteovenosa*). Ben is excited to find a Japanese umbrella pine (*Sciadopitys verticillata*) in the thicket, an endangered conifer that previously we've collected only from much further north. Its leathery, glossy leaves spring out in whorls from flimsy, fawn-coloured stems. To my delight, we encounter dozens more jack-in-the-pulpits growing among moss and ferns beside the track. Their fat, sturdy stems have a white bloom that catches the filtered light that percolates the canopy; one has a bright, lime-green spadix that looks fluorescent. We gather around a particularly robust one and Ben photographs me crouched next to it at the foot of the rocky cliff. Nearby, another interesting discovery lies in wait in the undergrowth: a feathery clubmoss (*Lycopodium* sp.) creeping out of the bank. Not a moss, in fact: it belongs to the most ancient group of vascular plants alive today, the first to have evolved roots, stems and leaves. Looking at its diminutive form, it is difficult to believe that 350 million years ago, its ancestors would have towered above our heads.

> *On the way glorious golden lilies mingled with the*
> *deep pink, varying to creamy white, of the mountain*
> *rhododendron, and the vermilion-leaved mountain ash, as*
> *we rose and passed along the sharp-crested granite ridge*
> *of Oya-shiradzu Ko-sliiradzu, whose flanks plunge steeply*
> *down, forest-clad, to the deep ravines of the Ojira-kawa*
> *and the Omu-kawa on either hand.*
> Walter Weston, 1918. [47]

We set to work, collecting snippets from all the plants we find and amassing them on the car bonnet. After half an hour the vehicle is littered with spidery branches and bits and pieces of

Arisaema iyoanum growing on Mount Kamegamori, Shikoku, Japan.

plant, as if someone has taken a machete to the canopy overhead. We process them all carefully, discarding them as we go. Ayako-san and Michiyo-san hold each plant up and inspect its features one by one, occasionally leafing through the large Japanese tomes to confirm their suspected identification, then nodding once each time they are validated. Ben scribbles down the name and we move onto the next specimen. With so many interesting plants to see, I find it hard to stay fixed to the spot, so scramble up and down the banks looking for new specimens.

One plant has evaded our scrutiny all day: a rare cobra lily (*Arisaema iyoanum*). The focus of today's work is to collect survey data, so all the plants we encounter – both common and rare – are important, but this is one I desperately want to see. It's related to the green jack-in-the-pulpits that seem to leap out at us desultorily all over the mountain. This rarer species, the books tell us, should grow here too, and as it is distinct there will be no mistaking it. Evidently more particular than its cousin, we'll need to search harder for this one. Fortunately our guide thinks she knows where it could be. We jolt and bump over the shoulder of the mountain in the four-by-four, deeper and deeper into nothingness. Untouched by civilisation, it feels like a timeless place.

We push the vehicle to its limits and then abandon it, setting off on foot down a green dappled track. It takes about 25 minutes to find what we think could be our plant: a specimen with an inflorescence long past its best, atop a stout, blue-tinted stalk that comes up to my knee. It's distinct from the green forms we've become blasé about since our first encounter with one this morning – that I do know. But we can't identify it reliably when withered. A handful more are jutting out from a steep bank, all in similar condition. I resign myself to the plants' truancy and saunter along the track, looking for other things to occupy myself. We're lucky to have seen it at all, I tell myself, kicking up the leaves. Ben and the others are chatting 20 metres behind me, pondering over a low-hanging sprig every now and again. I boot a fallen branch on the path and, to my left, something bright catches my eye. *Finally,* I smile. A thick blotchy stem, some 60 cm (2 ft) high and, at its pinnacle, a florid blossom, the size of a courgette. The cobra lily. I scamper up the bank to examine it. Its spathe is waxen, suffused with purple stripes that take on a blue, waxy hue at the base; the apex of the spathe is drooping over most of its length, so I lift it up like a flap and peer inside. It's the colour and texture of apple flesh, speckled with pomegranate at the edges. Its innards are chalky white, spattered with claret, and enclose a protruding spadix with colours to match. What an astonishingly beautiful plant. As I wait for the others to catch up, I spend a precious moment with it on this quiet mountainside in Japan's untrammelled wilderness, etching it on my memory; a moment that I know, when I'm home, will become set in paint.

> *The romantic beauty . . . for the most part, almost impossible adequately to describe.*
> Walter Weston, 1918.[48]

KYUSHU 九州

The Shinkansen (bullet train) deposits us in Kagoshima (鹿児島) on the southernmost main island of Kyushu, the 'Land of Fire and Water'. Kagoshima is the capital city of the Prefecture and has been dubbed the Naples of the Eastern World for reasons not immediately clear on arrival. The modern-looking, blocky city is a cauldron today and people have umbrellas up in the midday sun. We take a taxi to our hotel – a highly polished, imitation marble sort of a place, not far from the lazy grey Kotsuki river, and presided over by a Ferris wheel. After settling in I stroll along the paved bank of the Kotsuki, from which little stairways meander down to its muddy, grey beaches at intervals. Hot-looking residents in airy clothing are strolling quietly among the leafy cherry and red-flowered coral trees. The warm evening air is noticeably hazy, smoky even.

Not smoke, ash. Sakurajima (桜島) is Kagoshima's answer to Vesuvius (okay, *now* I see). Literally meaning 'Cherry Blossom Island', its gentle name belies the fury of Japan's most active volcano, which I now see looming behind the Kagoshima Bay. A fat blue cone, rearing out of the sea, its edges are smudged with grey, and strange, finger-like clouds hover over its crown. This geological monument is our destination for the next few days' botanical survey work. I'm excited by the prospect.

Excited and perhaps just a little uneasy. The volcanoes I've explored before now were all extinct. Not this one. I can see it smouldering even from over 10 km (6.2 miles) away. Is it safe to go? But Sakurajima's shadow fades from memory as I admire the miniature forests of ferns sprouting from the tree branches overhanging the river. They tell me I'm somewhere exotic; somewhere warm and wet, a place where unusual plants lie waiting, bursting out of every crevice. And I can't wait to find them.

Back at the hotel, I unpack my backpack, and shower to wash away the day. At my feet a trickle of water turns ever so slightly black.

ON THE FERRYBOAT from Kagoshima we read about the great eruption of 1914 on Sakurajima, the most powerful in Japanese twentieth-century history. After lying dormant for a century, deadly earthquakes and eruptions ripped across the island, spitting out vast eruption columns and lava flows. We debate whether we included pyroclastic flows in our 'field work risk assessment' (I've never been best in class at those). As we speak, dirty yellow clouds spill out of the summit and pour over the horizon. Signs are plastered all over the place to explain the island evacuation procedure in the case of an eruption.

We dock at the small port, which is framed by a scaffold of turquoise metal walkways, the sea smacking gently at its foot. We hire a car for the day from a shack at the side of the road and set off to explore the island in the hot, salty air. We soon spot the first plant of interest: native mistletoes (*Korthalsella japonica*) among the branches of small camellia trees planted along the side of the road. Presumably these have naturalised from native stands of camellia that grow on the evergreen forests which cloak the island. The grey-green, leafless branches resemble little bunches of rock samphire sprouting out of the trees, and aren't at all like the familiar Christmas mistletoes. We pass *minka* houses scattered along the coastal roadside, with sliding doors left open in the heat, and clipped *niwaki* trees in the gardens. Japanese politicians smile and give thumbs up from posters on every street corner; otherwise the island seems almost deserted. We make a right turn from the coastal road into the forest.

On the right of the verge, a lush thicket juts up from the concrete escarpment 2 metres above our heads; to our left, a bower of ragged banana leaves forms a shady overhead lattice. The vegetation is rather hard to make sense of because alien exotic vines festoon the native canopy. To our delight, we spot a chocolate vine (*Akebia quinata*), the fruits of which are eaten as a seasonal delicacy. The fat, star-like clusters of unripe green fruits splay out heavily on the palm of my hand. This roadside jungle-cum-garden gives way to a thick, bottle-green mantle of indigenous forest, forming a blanket over the hills. We take the first track into the forest we can find. To our astonishment, a few metres from the path, we see monumental vertical poles of

Japanese timber bamboo (*Phyllostachys bambusoides*) that had been concealed from the roadside by the dense vegetation. Their extraordinary glossy green stems are thicker than my arms and shoot up into the canopy. The track descends deeper into the murky forest. Here a profusion of contorted vines hang like a net from the canopy and little ferns and orchids poke out from tree trunks. A fig-relative we can't identify is pushing out its green, ball-like fruiting structures. We look up and down, then scribble down what we see. We discover the prize find on a rocky slope, is a stand of large *Arisaema ringens*. There are about a dozen, all with floppy, leathery leaves. One is clenching a fistful of green berries that bulge out of the withering brown spathe.

We head back to the car and explore the slopes near the port, noting officious signs stating, 'OFF LIMIT ZONE!' not far from an evacuation shelter. We suddenly realise that a film of pale grey ash has settled quietly over the island, covering its vegetation. As we return to the boat we notice it's also settled over us.

IT'S A BALMY evening, and everybody is eating al fresco back in Kagoshima. From the quiet purposefulness of the day, a vibrant cacophony of sizzling *yakitori* and *kanpais* has emerged among the red and yellow street lanterns. At a small street-side *izakaya* we quench our thirst with beer, and, with our chopsticks, pick at *torisashi* raw chicken (when in Naples!). All around us, groups of young Japanese men in matching white shirts and black trousers are taking shots of saké. We attract an attentive audience of locals in their mid-twenties who seem curious about our botanical expedition here. 'Where in America are you from?' enquires one. 'Oxford,' we reply. 'Yes, yes, Oxoford-oh.' They nod, as if to say well, yes, I thought so. 'But why the volcano?' enquires a girl in fashionable black clothing. We explain our aim to collect data from different types of plant community across the island, including volcanic vegetation. She giggles behind her hand, then fixes her eyes on us more seriously. 'You *know* that large eruption will happen?' We're unsure how to react. Then, playfully, she adds 'but not tomorrow'; so we all laugh again and make a toast to our good health.

7 AM. ANOTHER SEARING hot day and not a cloud in the sky. Except one. One big grey cloud growing from underneath.

Back on the boat, teenagers in uniform huddle around a tiny television screen rigged to the wall. They're all wearing yellow hard hats. Out on the deck, as we approach the volcano, the sea air seems hazier than it did yesterday and laced with the smell of something other than salt, something like burnt eggs. We disembark and take a car to the other side of the island. Heading east, lush vines are engulfing everything in their path and pooling out onto the road. As we skirt the south of the island, we spot a narrow lake and stop briefly to assess the aquatic vegetation. A mass of giant, rhubarb-like *Tetrapanax papyrifer* makes the lake all but impenetrable, so we head further inland and snake into the hills. In the distance we hear a rumble of thunder. Perhaps a storm is brewing.

1 pm. We stop for lunch at a rural *izakaya*. After removing our shoes, we're led through a complicated series of rooms to a low

table with *tatami* seating, a television blaring in the background. The air conditioning is a welcome relief. The din suggests the *izakaya* is busy, but this is difficult to ascertain because each compartment is screened off. We both order what the hostess translates for us to be 'ass of chicken' and discuss the botanical findings of the morning over a beer. The chickens promptly arrive, their asses each served with a raw egg poured over them, which proves difficult to extricate with our chopsticks. We consult our notes and chart our route for the remainder of the day.

3 pm. Not far from the volcano's summit, we locate a remote track leading to the very heart of the indigenous forest. This seems like the perfect spot to complete our survey work on the island. Sakurajima's active Minami-dake peak has rained ash over everything. As if in a dream, we enter an enchanted silver forest. I shake a branch above our heads and watch a shower of cinders fall silently to the ground. Along the trackside, rows of fingered *Fatsia* leaves hold hands across tussocks of ashen grasses. Tangled vines hang from the branches, all floury-white, like stone statues. It's a mysterious place. After about a quarter of an hour, a humped stairway of roots to our left leads us deeper into the gloom where an enigmatic, tropical-sounding bird is calling. Here we discover a forsaken shrine, half mossed over and obliterated by the thicket. Long-forgotten, and smothered in vines and ash, it seems like an eerie premonition of abandonment. Back on the main track, we busily jot down all the plants we can find. Along the way, between the dusty branches, we catch glimpses of blue-forested hills pitching down to the sleepy Kagoshima Bay.

4 pm. BOOM.

Everything is awake. An almighty thunderclap rumbles up out of the very bowls of the earth and roars at the forest. The vibration sounds dangerously close. Hastily we sweep up our pens, notepads and cameras, and dash back along the track, kicking up ash as we go. I can sense the island's lifeblood surging beneath our feet. Somewhere in the forest a bird calls repetitively, oblivious to

Sakurajima erupting,
Kyushu, Japan.

the primeval forces at play. It sounds shrill, like a siren. My mouth fills with brine and my heart hammers my chest. The prospect of the dangerous eruption we'd dismissed – dared even – now seems alarmingly real. No time to think. Run.

4.10 pm. More rumbling as we reach the car. Quickly, we shake off the ash that now covers us from head to toe, jump in the vehicle and speed off.

4.15 pm. The sky is no longer blue; now it looks almost yellow. All around us storm clouds gather, only this storm was not sent from above. *We need to leave now.*

4.30 pm. From the boat back to the mainland, we watch a solid column of smoke pour up into the sky and wonder: *what will happen next?*

7.19 AM THE DAY AFTER. We were forewarned. On 16 June 2018, the morning after we left the volcano, an almighty explosion occurred on Sakurajima. The eruption sent up a plume rising 4.7 km (2.9 miles) above the crater, and pyroclastic flows running 1,300 m (4,265 ft) down the volcano's south-western slopes. Significant ash fell over all the city of Kagoshima.

Japan sits in the Pacific Ring of Fire and has scores of volcanoes, but none as deadly as Sakurajima. Scientists have predicted that another volcanic eruption like that of 1914 is possible in under three decades and could threaten the safety of tens of thousands of people. Since the centennial anniversary of this eruption, the volcano has become increasingly volatile. But for now, people live

*peacefully in the shadow of Sakurajima, as if her storms
were a part of the weather; unflinchingly they put up their
umbrellas when it rains stones.*

THE RYUKYU ARC 琉球弧

'Typhoon,' says the check-in assistant, smiling, pointing at a red
and black board of delayed flights. 'Perhaps, flight cancelled?' (Still
smiling sweetly.) Is she *kidding* me? We have *just* fled a volcanic
eruption. Are we a magnet for natural disasters? Nevertheless,
about an hour later our flight departs for Okinawa and all
seems well in the Land of Fire and Water. Now we're headed
for the Ryukyu Arc in the south of Japan; we are entering the
Paleotropical Kingdom.

Seatbelts on. We're warned to brace ourselves for turbulence;
nothing I haven't experienced before, I tell myself with confidence,
and settle down quietly to writing up our adventures on the
volcano. Americans are on the plane; it seems strange to see
Western faces again after so long. It's a short flight and before we
know it, we have 20 minutes until we're due to land.

Drama. A sudden jolt sends my pen to the floor where it
rolls out of sight. Having seemingly fallen out of the sky for
a second, the plane steadies and murmurs simmer around the
cabin, followed by snatches of nervous laughter. Ten seconds
later, a more violent lurch. No laughter this time. The turbulence
has tapped into some primal fear, and a surge of adrenaline is
released across the cabin. Someone gasps. It's OK though, I
remind myself, as we're tossed violently from left to right, because
turbulence alone can't take down a plane – I know, because I
read it somewhere. (Didn't I?). Even so, there's something more
than a little disturbing about this violent lurching and dropping,
and this is a tropical typhoon, after all. But three jerky attempts

to land later, we finally bumble and bump down onto the runway of Naha Airport. With shaky legs we disembark, all the while slammed by gusts of a stiff, wet wind. I feel giddy. The small part of me that likes a drama finds this thrilling – landing in a typhoon – a story for the pub back home. And it's just the warm-up for a tropical adventure, I can tell.

How troublesome and dangerous the voyage to Japan is, and how boisterous and subject to gales the sea is . . . Thunder is by no means unfrequent; but tempests and hurricanes are very common, as also earthquakes.
C. P. Thunberg, 1775.[49]

WE CAN'T RESIST stopping off at a tropical-looking beach to stretch our legs. White sand. Jade blue sea. A leafy panorama delineated from the sky by a ribbon of blue-grey hills. These dreamy views are defiled by a swarm of overhead electric cables writhing all over Okinawa. Trying our best to ignore these, we notice the sharp white sand is littered with fossil-like remnants of mushroom coral and sand dollars. We wade into the warm, shallow water where jet-black sea cucumbers spatter the rocks. I stroke one. It feels rough and warty like a toad. I venture a little deeper, pushing down into the seawater that floods my sinuses. Two metres below the surface I find a blue linckia starfish (*Linckia laevigata*) the size of a dinner plate and the colour of stormy skies. Back on land, at the far end of the beach, a nose of rock juts into the sea bristling with palm-like, sword-shaped

A blue linckia starfish
(*Linckia laevigata*) in
Okinawa, Japan.

leaves. On closer inspection I discover these belong to a patch of screwpines (*Pandanus tectorius*). To my delight, a handful are in fruit. They hang heavily among the papery bunches of leaves like green and orange pineapples the size of rugby balls. The strong branches of screwpine are interwoven with leafy panicles of beach hibiscus (*Hibiscus tiliaceus*) scattered with rhubarb and custard-coloured flowers. I pause on the outcrop to enjoy the plants with warm sun on my skin, as the sea slaps the wet sand lazily a few metres beneath me. We stroll for an hour, happily acquainting ourselves, plant by plant, with this new realm. The prospect of a week in this tropical paradise feels as glittery as the sea stretching out before us.

OUR ACCOMMODATION IS a throwback to the 1970s, complete with pink and apricot sanitaryware and shag carpets. More importantly, I discover gleefully, this gaudy hotel and its neat little golf course are nestled in hills thickly blanketed in subtropical rainforest. In a cloud of insect repellent, I cross the eye-searingly pink air-conditioned threshold to explore the warm, wet wilderness.

The first plant of interest I encounter at the edge of the forest is shell ginger (*Alpinia zerumbet*), a plant I know well from a life spent poking about tropical hothouses. A 'multipurpose plant', shell ginger has all sorts of uses; here in Okinawa its leaves are used to make herbal tea and for wrapping rice. Its lush, ginger-like leaves flop from rigid stems that poke out vertically from the undergrowth. About a dozen have sprays of pinky-white and yellow flowers. They remind me of suspended shoals of exotic shrimp. Nearby, on a ferny bank, enormous, paddle-shaped leaves of giant taro (*Alocasia odora*) thrust out upon fat, watery stems. I tap one of the stout stems and its elastic sheet makes a satisfying thunder-like wobble above my head. Close to the ground, a handful are in flower. Each of their curious shapely spathes, the size of fat bananas, encloses a cream-coloured, finger-like spadix teeming with tiny black insects. Above my head, the fork of a meandering red tree trunk is crowned with a giant bird's-nest fern (*Asplenium nidus*), throwing out shiny, curly-edged fronds the colour of apples. This excites me most of all, which is hard to explain, because it's common here in the tropics. Maybe that's why I like it: as an emblem of a rainforest – a place where countless plants grow to the very heartbeat of life on earth.

WE MEET ATSUSHI-SAN, Director of the Botanical Laboratory at the Okinawa Churashima Foundation, and Kensei-san, who is to be our guide. We sip iced green tea around a white metal table while they tell us about the mission of the Foundation to lead research and public awareness about Okinawa's rich subtropical flora and fauna. Atsushi-san asks me what my interests are specifically. Parasitic plants, I tell him. He nods once and blinks slowly, as if to say 'that is correct,' then wanders off to find something. He returns with a large, clip-top preserve jar full of liquid the colour of urine. In it bobs a clotted brown specimen that looks like the spore-producing structures of a horsetail plant – a clutch of fat, finger-like sprouts. *Tsuchitorimochi*, he says, smiling. I pick up the jar and peer at the curious plant as it slowly sinks to one side. I know just what it is: *Balanophora* – a forest leech of a plant that sucks sap from the roots of trees. I love it! Next, we're taken backstage to the lab where we're shown row upon row of conical flasks containing micro-propagated plants. Here we're introduced to a student – a woman in her early twenties who, with Kensei-san, will join us in the field. Atsushi-San wishes us luck with our adventures, and we set off to explore the jungle.

AFTER A SHORT drive, we stop to examine a snatch of rainforest Kensei-san knows well that lies close to the main road. We pile out of the car and look up at a cathedral of trees towering above us, from which roots and vines dribble down from to the floor. Beneath the canopy, the tops of palms splay out from clouds of shrubbery as if floating; at the foot of the trees, unfolded ranks of giant taro leaves mirror one another, as neatly as if they were planted. I'm reminded of the Palm House at Kew. Kensei-san leans against the car, smoking, while the three of us bring him bits and pieces of plant we're unsure of to identify – grasses, leafy twigs and so on. Curious, blue-brown, snake-like fruits coil down from a tree branch – I don't know what these are, and nor does Kensei-san, so I make notes in order that we might identify them later (we never do though). The place is deliciously rich in plants. Our first taste of the subtropical rainforests of Okinawa, and I'm hungry for more.

Our next task is to collect seeds from a tropical vine called *Aristolochia liukiuensis* for the tropical plant collection at

Oxford Botanic Garden. It's common here, Kensei-san tells us, and sure enough, we see plenty of it from the car. We park on a steeply inclined back road and wander down it, searching for a fruiting specimen. They're easy to spot coiling around slender tree trunks, sending out vertical rows of shiny, horizontal, heart-shaped leaves, each facing palm-up. Eventually we spot one with half a dozen green, pear-like fruits hanging a couple of metres above our heads. We use pole-cutters to sever one from the vine and it thuds onto the road unceremoniously. Kensei-san peers at it, seeming a little unimpressed – perhaps he thinks it's unripe. 'Two fruits,' he says, fiddling with his lighter, 'two fruits'. So we lop off another one.

Half a kilometre down the road, to my delight, we see a stand of flying spider-monkey tree ferns (*Cyathea lepifera*). Five are growing together, all at different heights with bent, slender trunks winding their way up to the sky, each topped with a disproportionately large ring of feather-like fronds. They look prehistoric – as if they belong more to a time of long-necked dinosaurs than our own – which in a way they do. Then, as if hearing my silent musings on prehistoric plants, Kensei-san points out another one growing on the verge – this time a cycad (*Cycas revoluta*). I push my way into the thicket to get a better look, checking around my feet for deadly snakes, as Kensei-san has warned me to. The cycad is a glorious-looking thing. It has a bristly whorl of bottle-green leaves that shine blue in the light and from its centre projects a yellow, pollen-bearing cone, like a gigantic corn on the cob. We see a few other noteworthy plants nearby, including the dye-yielding red kamala tree (*Mallotus philippensis*) and the medicinal, pink-flowered Indian rhododendron (*Melastoma malabathricum*). Just as we decide to head off for lunch, we notice movement on the forest floor. A patch of rotting fruit is writhing with beetles and little hermit coconut crabs, which seems surprising as the ocean is miles away. Kensei-san explains that these are terrestrial crabs. Dozens of them scuttle about like spiders. 'Kawaii!' cries the student, raising up her arms – 'too cute, too cute!' She crouches to pick one up and it retreats into its shell, with two beige claws just protruding. Then she screams. The creature has taken offence to the intrusion, jutted out of its shell and pinched her. It dangles from her hand which she is holding out in despair, while looking at us helplessly and yelping in pain. It won't budge and we're all at a loss to know what to do. Finally, Kensei-san pours a bottle of water over the situation as if her hand is on fire – which from her face it might as well be. To our relief, the crab then thumps to the floor, leaving in its wake a gaping wound and sorry-looking student. We head to an ocean restaurant to see if we can console her with a good Chicken Katsu. 'Well,' mumbles Kensei-san through the cigarette bobbing between his teeth, while looking at his watch, 'it solves most things.'

In the afternoon we head for the Yambaru National Park (やんばる 国立公園), the island's vastest tract of pristine subtropical rainforest that covers its northern quarter. On the way we stop at Kunigami and stare out at the vast North Pacific. Great, grey, forest-crested cliffs drop deeply onto the yellow sand; beyond them stretches mile upon mile of dimpled blue ocean. I say mile: 7,600 miles to be precise, for the nearest mainland in front of us is 12,230 km away in Mexico. On the edge of the cliff perch little shrubby cushions of Indian hawthorn (*Rhaphiolepis indica*), which must be able to withstand a good pinch of salt. We drive on to a notch of yellow sand along the wild northern coastline to find the endemic Bonin Island juniper (*Juniperus taxifolia*) that Ben is keen to see. Kensei-san knows where it is, but without climbing gear, collecting seed is all but impossible (even for us!), for the plant is growing eight metres up a sheer cliff.

We drive past pine forests and sedges growing on orange clay and park at a remote stretch of coastline. Down on the beach, thick trunks of screw pine arch out of the coastal thicket like raised elephants' trunks; from them hang dozens of waxy-orange lumpy fruits, sculpted with hexagonal buttons. A stone's throw from the surf, sprawling yellow, spaghetti-like stems of dodder (*Cuscuta hygrophilae*) appear. This parasite wraps itself around the surrounding vegetation from which it steals its food via little syphon-like plugs called haustoria. As we push our feet along the sand, we spot coastal inlets flanked by tall cliffs with tropical vegetation tumbling over the edge. At their foot, morning glories (*Ipomoea pes-caprae*) creep across the sand, their rose-pink flowers already melting under the hot afternoon sun. Their fan of stems converge where a clump of giant crinum lilies (*Crinum asiaticum*) are jutting out, improbably lush in the dry sand. One has a ball of pristine white, spidery flowers as large as a melon. Most curious of all, as we return to the car the student spots a patch of sea mangos (*Cerbera manghas*) sprouting out of bare sand among some old lumps of bleached coral. They look for all the world like miniature coconuts, with stout stems pushing assertively from fibrous brown balls. One has glorious white flowers like those of a periwinkle; but its beauty belies its deadly poison, hence its other more macabre name of 'suicide apple'.

Dear Diary.
I've been poisoned: sick, sick, sick is all that I feel; sip
water and stare at the wall is all I can do; and round
and round and round goes the wall in a godawful dizzy
blur. My heart beats quickly and a cold sweat creeps
over me. I think I might die here.

TWELVE HOURS EARLIER ...

Our hosts have organised a celebratory dinner for us in
Nago. We file into an *izakaya* with dark, wood-panelled walls
plastered in placards of food, drink and toothy smiles, and the
smell of fried chicken mingled with cigarette smoke wafts in on
a warm breeze from outside. We nod politely to one another
and arrange ourselves around the wooden table. We're joined
by our hosts, Atsushi-san and Kensei-san, the student, and
a loquacious middle-aged lady who tells us she's a bestselling
author. She speaks good English and is inquisitive about the
conservation work we're doing here. We all chat companionably
about the wonderful plants we've seen in Japan as rows of little
white bowls accumulate in front of us. Goya Champuru is on the
menu. This local stir fry dish has melon, tofu, egg and pork, all
served with steamed rice and miso soup. *Itadakimasu*, we chorus
in synchrony, to express our thanks.

We wash the Goya Champuru down with lashings of Shōchū,
an earthy tasting liquor which is Okinawa's answer to saké, only
a tad stronger (as we shall see). Shōchū confers many health
benefits, one of our hosts explains sagely, translating the bottle's
label and then illustrating his point by tapping various parts of his
anatomy. I'm a few glasses in, and certainly feel good on it. Never
better actually: it does exactly what it says on the bottle. And
everyone seems impressed by the amount I'm quaffing too, which
is nice – I must be doing untold good to my insides.

I'm a bit giddy though. No, merry. Nothing wrong with merry.
And what's this? Another drink? Because we've enjoyed the
Shōchū so much, a prized liquor called Shima or 'island saké' has
been ordered especially for Ben and me, as honoured guests. It's
brought to us in an elegant dark glass bottle with spidery Japanese
characters on the label, no doubt spelling out manifold health
virtues. 'Gosh, this is nice,' we say, as everyone nods approvingly.

Gosh, this is strong, I think, as I reach for my glass unsteadily. I
really don't need another drink, but the Shima *has* been ordered as

a gift and it *does* look expensive. Better have one more – you have to be polite, don't you? It doesn't help that it's considered *impolite* to pour your own, so my intake is completely at the mercy of my hosts. I'll sleep it off.

Sleep sounds nice, actually . . . anyway, it's good for me, right? We stand to exchange gifts and speeches. I talk effusively about how much I love Japan, its plants and its people. And I mean it, too. But careful not to knock your glass over. Am I swaying? The wooden walls make me feel like I'm on a boat. I sit back down, narrowly avoiding missing the edge of my chair. We all laugh hysterically. 'How does . . . how does your hand fffeel after that hermit crrrab bit you?' I ask the student. Have I asked her this already? Ah yes, I think so. Well, it looked painful, that I do know.

As I look around the room starts to spin. Faces on posters smile and wink – round and round they go. I can't fathom how on earth I've become so soul-baringly drunk on this 'health tonic', or how this is going to end. 'Ben, I'm . . . *ab* . . . absolutely *gone*,' I whisper loudly in his ear. 'Out of my tree,' he agrees, and slides slowly down his chair.

Dear Diary.
Nothing to report today.

KENSEI-SAN SHOWS us around the Okinawa Churashima Foundation's plant collections grown under glass. Roots and vines tumble out of terracotta pots in long rows upon metal tables. The collection contains rare species collected from remote tracts of rainforest all over the island. We part the wiry stems of an asarum (*Asarum gelasinum*) and peer at its curious, three-parted flower; it's the colour and texture of a beefsteak. Next we're shown a species of aristolochia we've not seen before (*Aristolochia zollingeriana*), which possess a tangle of vines throwing out curvaceous, tubular flowers the size of my little finger. A predominantly South-East Asian species, this is the northernmost outpost of its range. Kensei-san plucks a fruit for us so that we may grow it back in Oxford for the first time.

We drive along the coastal road for half an hour, then head inland to Mount Katsuudake. At 414 m (1,358 ft) Katsuudake is not a difficult mountain to conquer, but it's 31°C (88°F) today and very humid. And we're a little tired now, after weeks of trekking. We prepare ourselves for the hike – Ben and I with insect repellent and suncream, and Kensei-san with a cigarette. The trail begins with broad steps lined with drippy ferns and branches. The blocky stairway peters out into a lumpy, ill-defined path of knotted roots clamped to the jagged rocky slabs. After about 20 minutes, Kensei-san points out a stand of wild citrus (*Citrus depressa*) known as 'shīkuwāsā'. The lanky trees are about shoulder height and scattered with little greenish-orange fruits like satsumas.

Kensei-san says there are several species of orchid we should see on the ascent to the summit but we can't find any, despite searching high and low. This is a little disappointing, for the orchids of Okinawa are splendid and we haven't seen one yet. But close to the path we find rope-like vines of a waxvine (*Hoya carnosa*) – a popular house plant the world over – and seeing it grow wild makes up for the absence of orchids. Out of its tangle of stems point thick, dark green leaves flecked with tiny white dots, like stars in an emerald sky. Next to the waxvine sits a strange snail the colour of grey and black marble with a row of little spines spiralled around its shell. I've not seen one like that before. We press on. A crawling web of roots converges, leading the eye up pillar-like trunks to a leafy ceiling, from which vines drop back down to the ground.

The approach to the summit entails a short scramble up craggy ledges. The trees thin out as the rocky crown of the mountain appears above us, and a welcome breeze whips up. Out on the broad-topped plateau we rest, and sit on wrinkled, blocky boulders the colour of unpolished silver. It's calm up here. There isn't a sound. Grasses shimmer. Black and yellow butterflies flit around our feet, then bounce silently away on the breeze. My eye follows one out onto the horizon, over softly forested domes that rise and fall out into an ocean of cotton wool clouds. As the creature vanishes, I notice in the distance a grey rash of quarries, tower blocks, houses and pylons. How different this view must have been for those who wandered over these rocks just a few generations before my own and looked out onto a landscape unspoiled – just vegetation and glittering sea, in one of the most biodiverse parts of all Japan. The trees remember. Forests the world over speak the language of loss. But I don't want to think about that today. I turn the other way and look out instead at Japan's living, breathing hills that fade green to blue as they roll into the sea, just as they always have, and think about the journey.

Months and days are the wayfarers of a hundred generations, the years, too, are goings and comings of wanderers . . . I . . . drawn by a cloud wisp wind, have been unable to stop thoughts of rambling.
'The Narrow Road to the Deep North', Matsuo Bashō, 1644–94.[50]

7

TO PITCHER PLANT
PARADISE

MOUNT KINABALU, BORNEO

7

TO PITCHER PLANT PARADISE

MOUNT KINABALU, BORNEO

They came true, those dreams. Dreams of a boy who longed to become a botanist when he grew up, to scribble and sketch his way around the world, whatever it took.

IMAGINE A MOUNTAIN reputed to have more species of fern than the whole of Africa, a place dripping with orchids – hundreds of different species – and the richest assortment of tropical pitcher plants conceivable, five of which are found nowhere else on earth. Such a place exists: it is Mount Kinabalu in Sabah, Malaysian Borneo, and it's a botanical wonderland. Standing 4,101 m (13,455 ft) tall, the gargantuan mountain is the largest in the Malay Archipelago and the twentieth most prominent in the world. Its jagged black pinnacles jut out from a collar of cloud veiling its imposing outline. As a child I'd stare at photographs of the mountain in books and dream about climbing it. From the other side of the planet it haunted me.

Botanists have always been captivated by the place. Lilian Gibbs, a British botanist who became the first woman to ascend Mount Kinabalu in 1910, wrote an account of the mountain's flora in 1913. She describes how local people 'invested it with a wealth of legendary lore . . . On its mist-crowned summit the souls of the departed find their eternal home, phantom herds of buffalo follow their masters to graze on the shadowy grasses which abound in that fabled kingdom'.[51] Its very name is reputed to mean 'the revered place of the dead'. Seeing the mountain looming on the horizon, it's easy to appreciate how it became as much shrouded in legend as by mists.

Kinabalu became a place of pilgrimage for botanists who, over the centuries, left a rich legacy from their collections and accounts of their adventures. Its mosaic of black, slippery slopes, mossy outcrops and wet, ferny forests are home to a bewildering diversity of plants. Many have been observed or collected only once. The eminent botanist Professor Edred Corner once described the mountain's flora as 'the richest and most remarkable assemblage of plants in the world'.[52] The first documented ascent of Kinabalu was made by the botanist Sir Hugh Low in 1851, at that time a British Colonial Secretary. Low was one of an adventurous group of nineteenth-century European plant collectors whose expeditions contributed significantly to the advancement of plant science. Kinabalu's highest peak and a deep gully, dubbed one of the least explored and most inhospitable places on earth, are both named after him. Low collected the 'king of pitcher plants' (*Nepenthes rajah*) on his expedition in 1858, described the following year by Sir Joseph Hooker as 'one of the most striking vegetable productions hither-to discovered'.[53] To this day the most spectacular species of pitcher plant described (to me at least), it produces pitchers the size of house cats. Who wouldn't be entranced by that? Low made three visits to Mount Kinabalu; two with Spenser St. John, the Consul General of Brunei. St. John recorded lively accounts of their adventures which featured the pitcher plants they came across, some of which I've included in this chapter.

Interest in the mountain continued into the twentieth century. Mary Strong Clemens was a passionate botanist who went to the wildest reaches of South-East Asia to collect plants. She and her husband Joseph spent many months collecting on Kinabalu in 1915, and again from 1931 to 1933, leaving the mountain only briefly to identify and sort their collections in Bogor, Indonesia. During their two years in residence, spent in makeshift lodges and tents, they amassed a collection of many thousands of plants now stored in herbaria around the world. To this day, many of the specimens have not been examined in detail.

I was inspired by these botanists' adventures and started planning my own. I could feel the mountain drawing me towards it, like a magnet. In 2005 I travelled around Borneo and immersed myself in Kinabalu's flora. That summer was a green blur. I relished every leafy square metre of the place, consulted every

The *Nepenthes* pitcher plants of Mount Kinabalu, Borneo.
Clockwise from top left: N. x *kinabaluensis*, N. *burbidgeae*,
N. *fusca*, N. *rajah*, N. x *alisaputrana*, N. *lowii*.

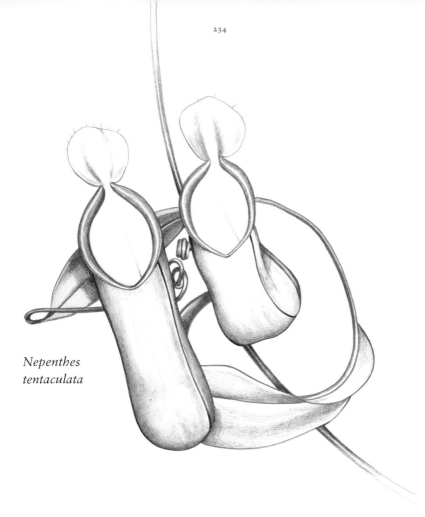

*Nepenthes
tentaculata*

orchid and pitcher plant, and scribbled them all down like a man possessed. I climbed up and down that great mountain twice in one week, once by the popular trail from the Timpohon Gate and once by the harder ascent from Mesilau, a trail now closed indefinitely due to its inaccessible paths. Every footstep up that mountain fed my plant addiction; it nurtured a seed, planted in my teens, and set something in motion. I slithered and scrambled my slippery way over rocks, up trees, down ropes and into the very heart of pitcher plant paradise.

MY DRIVER IS a 23-year-old Malaysian man called Walter Maximus. He tells me he's a fellow pitcher plant enthusiast. Can you believe it? 'You are going to love it here,' he says, laughing. We're accompanied by two brothers from West Malaysia, Ken and Sean, and we all make polite conversation and get to know one another as we snake up into the hills in the old diesel-drenched four-by-four. We pass little shacks clinging to the cliffs and stalls selling tropical fruits, all piled high, strung up and rolling along the road. The air becomes fresher and ferny banks rise steeply from the road. The immense mountain looms into view suddenly. At first it's like a bruised cloud on the horizon, then an unmistakable rugged mass: a great dinosaur rearing up out of the rainforest, with leafy ribbons clawing at its feet. This is it – the place I dreamed of. My heartbeat quickens.

1,866 m (6,122 ft). We register at headquarters and drive to the Timpohon Gate to begin the ascent. I'm impatient and twitchy – like an athlete before the race. We meet our guide at the gate, a small Malaysian man in his forties with a serious demeanour and muscly brown legs. We hoist up our rucksacks and set off into the forest. It's a tropical lower montane forest, dominated by oaks (*Lithocarpus*), chestnuts (*Castanopsis*) and the occasional conifer (*Podocarpus imbricatus*). It has a rich earthy smell to it, like moss and decay – it feels like a British woodland in October. We pass weary-looking climbers finishing their descent from the summit; none of them are smiling. After about a kilometre we reach Carson's Fall, one of the mountain's many waterfalls, dribbling and spitting over the black rock. A profusion of mosses and ferns bathes at its foamy mouth. The first 6 km (4 miles) of the 14-km (9-mile) trail are not too arduous and there are plenty of opportunities to stop and admire the flora. There isn't time to scrutinise them all, but I pause regularly to observe the most eye-catching, such as the flopping, hand-like fronds of *Dipteris conjugata* reaching out at our feet.

Nepenthes x *kinabaluensis*
on Mount Kinabalu's
Summit Trail, Borneo.

You never forget your first *Nepenthes*. To a plant addict I guess it feels like winning at football. My first is *N. tentaculata*. It has five perfectly-formed purple pitchers pushing their way out of the moss. I'm not sure the others in the group are as excited as me, to be honest – they seem keen to press on, as we have a long climb ahead. Luckily we find dozens more, further up. Some have formed vines that creep along branches of rhododendrons (*R. stenophyllum*) and dangle from the branches above our heads. The largest are the size of my thumb and yellow – the colour of starfruits – diffused with red, like spray paint. Their gaping, diamond-shaped mouths are pale, like the bloom on a grape. I tap one gently and watch it sway daintily on its tendril. It's so beautiful. I feel a buzz run through my fingers, like a drug. Often and often I've dreamed of this moment, my first encounter with a pitcher plant. I've waited years for it; my whole life, in fact. Now it's arrived and I feel like dancing up the mountain to find my next one.

2,252 m (7,388 ft). *N. tentaculata* was just the warm-up, I'm about to discover. The path steepens as we meander up to the third shelter, Pondok Lowii. Views of the concertinaed collar of the mountain flicker between the trees, shifting through the clouds. The air is fresh now and the vegetation is sparser and more stunted. The soil becomes a yellow, muddy clay, the same colour as the rock. This sets the stage for the next *Nepenthes*. We find it in the clouds, a grove of *N. villosa*. Dozens sprawl in and out of the reedy, fern-packed ditches, throwing their bronze, toothy pitchers all over mountainside, like melons on a vine. Many are out of reach, frustratingly, but I find a few nestled beside the soggy track. I kneel to examine one and run my fingers along its rows of red teeth. The rest of the party seem happy, now they can see how much it means to me – can see it written all over my daft, awestruck face! Soon they're pointing them out faster than I can look at them. A friendly competitiveness to find the biggest one unfolds on the mountain.

2,900 m (9,514 ft). It's so cool I've put my coat on. It's hard to believe I started the day in the steamy tropics. I'm buzzing from all the plants I've seen but can feel a slight strain in my calves coming on. Our guide takes me to one side, leaving the rest of the party on the main track. He blinks and tells me there's something I'll like nearby, 'something good'. We push into a scratchy thicket dominated by conifers (*Dacrydium gibbsiae*). It smells like wet terracotta. We reach a clayey hillock covered in lush ferns with tussocks of sedges among the intricate branches. Then I see it: *N.* x *kinabaluensis*. I stand there, gawping. I can discern about 15 plants in the wet gloom among the trees. Most of the pitchers are brown and papery, but a handful are fresh and blood-coloured, with large lids and ruby-red peristomes. They

have a faint pleasing smell that I find hard to place – rather like old books. I peer into one and see the remains of drowned insects sloshing around the base. Another one appears to have opened this very day. It has a pristine ruby peristome and smooth, glossy yellow interior. I could spend hours with these plants, I really could. But we need to press on to our resthouse before the worst of the cloud sets in.

Soon the serpentine rock becomes dominated by a miniature forest of contorted, grey-stemmed *Leptospermum recurvum*. They look like Japanese Niwaki trees draped with pea-green conifer chandeliers (*Dacrydium gibbsiae*). Oddly, the marshy, hummocky slope has the feel of a moorland, only lusher. *Nepenthes villosa* is common here, and I find pitchers framed beautifully by the white and yellow feathers of a small orchid – probably *Dendrochilum stachyodes*. The climb becomes more tiring as we approach the Pondok Villosa shelter, and we pause regularly for short breaks. My skin feels cool and clammy. I put my woolly hat on, as much to keep the water from dripping into my eyes as to keep warm. Between the gnarled trees I catch glimpses of vertical cliff faces, clothed in an olive green mantle with clinging white clouds. The blocky orange path spirals its way up through clumps of reeds and ferns beneath a canopy of upturned umbrella-shaped *Leptospermum* trees. A little way from the track, an exciting find is a wild rhododendron (*R. lowii*). It has large, lax bunches of yellow, trumpet-shaped blossoms like puddles of sunlight in cloud. Soon we find another species (*R. fallacinum*) with loose balls of scarlet flowers. It's like a mountain garden that planted itself.

Nepenthes villosa on Mount Kinabalu's Mesilau Trail, Borneo.

3,000 m (9,842 ft). Things are changing rapidly. Both the vegetation and the air are thinning. We pass a climber sitting by the path, complaining of a headache. Vast naked rock faces jut out above us, their slaty-black mass looms ominously. Fast-moving white mists pour and swirl around grey-green forested terraces. Soon, but for brief glimpses, views beyond 10 metres away all but evaporate. We can scarcely even see each other at times. As we walk through the shifting clouds, bald patches appear on the mountainside – vast domed slabs of granite. Cushions of soggy, spongy moss spring up at our feet. The rough, ochre-coloured track skirts bonsai-like trees draped in mosses and lichens, dripping white-flowered orchids and goodness knows what other epiphytes. Tall sedges (*Machaerina falcata*) stand rigidly out of the seepages. *Coelogyne papillosa* orchids form clumps of pleated leaves like aspidistras; breathtakingly beautiful white flowers arch out from them. I admire them, dangling daintily from their slender orange pedicels.

3,272 m (10,735 ft). We reach Panalaban and the Laban Rata resthouse where we'll spend the night. It's an unattractive cream building with rows of small windows and flimsy-looking ladders and terraces bolted on. Elephantine walls of summit rock rise steeply above its roof. Inside, a throng of people have congregated and the air is thick with weary content and menthol muscle rub vapour. What I'd give for a cold beer. But beers are not on the menu (not much is), so I pick at my buffet meal of noodles and brownish vegetables. I sleep for just one hour under a scratchy brown blanket. I swot small brown cockroaches, but they seem to pour into the bed from every direction. Around me, I hear whispering about travel itineraries and turtle sanctuaries. I stare at the roof of my bunk bed and watch pitcher plants and orchids unfold in the dark. At 2 am we set off for the summit.

Even paradise is afflicted by typhoons. Although considered a very conquerable mountain, Kinabalu can quickly become treacherous in bad weather. We form a line by torchlight, and proceed up the steep summit trail with grim determination. I feel faintly nauseous from the altitude (perhaps it's just as well the Laban Rata doesn't serve beer). Cold wet blasts of typhoon pin us to the rock in unison. This wasn't what I had in mind when I'd pictured myself ambling among the plants. Still, no going back now, I think, as I spot a handful of other climbers opting to do exactly this. With white, wet hands, I soldier on with the others. We clamber wet-shod up steep, slippery wooden ladders and over cold, salt and pepper-coloured rounded boulders, supported by guide ropes. In the lees of these blocky walls I see stunted silvery thickets peeking out of the crevices, but visibility is poor and there's little chance to botanise.

3,668 m (12,034 ft). We reach the Sayat Sayat checkpoint at the foot of the summit in howling wind. Sheets of rain rise from beneath us. Waterfalls form at our feet. It's too dangerous even to attempt Low's Peak, so we gather in the shelter, cold and wet, waiting for it to pass. By daybreak the typhoon calms to something like a bad day in Snowdonia, and we pick our way back down to the Laban Rata for a warming cup of 'milo' (hot chocolate). Along the way, the cloud disperses to reveal strips of the highest dwelling rhododendron on the mountain, *R. ericoides*. It looks like heather stuffing the cracks along the summit plateau. Many plants that grow here are notable for their affinity to mountain floras elsewhere in the world. I could spend days studying them, working it all out.

Our descent through the storm-washed scrub is eerily quiet in the hanging mists and fog. It's a lonely place. I'm mindful of the people who have lost their way on the mountain, some never to be seen again. The occasional frog pierces the damp, sound-muffled air with a loud croak. Frustratingly the strap on my rucksack snaps, so our resourceful guide binds it with an elastic band (that holds fast for a month). Back at headquarters I eat sweetcorn soup and rice while waiting for the bus, lamenting finishing it before I've even started. My legs hurt. I wonder when I'll be ready to climb again.

That night I watch tree shrews dart among the rainforest branches at sunset as I sip jasmine tea, contemplating the beautiful plants I – like all the botanists before me – have encountered.

To see these plants in all their health and vigour was a sensation I shall never forget – one of those which we experience but rarely in a whole lifetime!
F. W. Burbidge, 1880.[54]

MY SECOND ASCENT of the week is by a longer trail. I set off
early, with a middle-aged, jovial driver. She chants Malaysian
expressions, then laughs at me affectionately as I try to imitate
them. As the vehicle chugs its way into the hills, we pass steep,
streamed banks of ferns and grasses, and the air grows cooler. We
stop at Pekan Nabalu, a viewing point with market stands near
the foothills of the mountain. Tables are piled high with green and
brown pyramids of durian fruit that fill the air with a gassy odour.
They look like sea urchins. There are two kinds sold here: one is
short-spined and green, the other is called 'jungle durian' and has
long, fierce spines. It's as large and round as a football. I buy a
large brown one for RM 4 (four ringgit, the Malaysian currency).

The woman deftly cuts it open, and I peer at the slippery, stringy flesh in which large, almond-like seeds glint. The custard-coloured fruit has the consistency of a ripe avocado, and tastes like boiled sweets, only with a good dose of raw garlic. Odd. It warms my insides though, like a dram of whiskey. Something for the road.

In the spitting rain, I wait at Mesilau for my guide. After half an hour he appears. Ramin is a slim man in his late thirties, dressed in red mountain gear. He tells me proudly that he had a baby boy the week before. I pat him on the back, and we set off over the vast eastern ridge of the mountain. The grey, rocky path is bound by a web of buttress roots. But for our footsteps and the occasional babbling stream, the moss-muffled forest is silent. The wet air smells sugary, like autumn leaves. Steeper paths have scales of rickety wooden steps, flanked on either side by sheer banks. Above them, a web of rotting stumps and branches is blanketed thickly with rust-coloured moss, from which bracken-like fronds arch out. The steps give way to a little orange path, carved into the rock, which ducks and dives in and out of little mossy woodland glades containing white and yellow orchids. 'Sam sam,' exclaims Ramin, pointing at a patch of little begonias. We stare at them. But there isn't time to identify the plants here, so early in the climb, so we push on. Soon I encounter the finest *N. tentaculata* I've seen yet. It has squat purple pitchers at least seven centimetres (3 inches) long, half pulled into the moss, and larger pitchers the colour of ripening lemons, conveniently dangling from the branches at eye-level. Soon we see dozens of its delicate stems weaving in and out of the scribble.

We pause and Ramin offers me some little snacks his wife has made for him, little chewy balls that taste strongly of peanuts. We sit quietly, enjoying the fresh mountain air. He's a man of few words (to me, anyway), so it's a surprise when he asks me what I plan to do with my life. He can see I'm interested in plants, that's obvious. He probably thinks I'm on some nature kick. So I don't tell him I plan to dedicate my life to plants; that I'm obsessed with them, and plan to chase them around the world. He'd think I'm unhinged, wouldn't he? Most people probably tell him about a career in IT or something like that. I mumble something about ecology and conservation and he seems satisfied with that.

Our quiet progress over the mountain's eastern shoulder is punctuated by pauses, every half a kilometre (one-third of a mile), at little hexagonal wooden shelters. By the first, Pondok Schima, views of blue-green fluted cliffs glimmer through the open tree canopy. Intricate plumes of a dark green fern (*Monachosorum subdigitatum*) bow overhead. Moss creeps up into the trees and hangs like soft stalactites from the branches. I touch one and it discharges a stream of water down my arm. From Pondok Bambu, we descend back down into the blue shade of the underbrush to the auspiciously-named Pondok Nepenthes. We cross the Kipuyut hanging bridge, a contraption of wooden planks and green, tennis court-like nets, walled on either end by reeds. Beneath us, the West Mesilau river sloshes around smooth, glossy boulders. After ascending little wooden ladders into dappled sage-coloured forests, we emerge from Pondok Lompoyou onto a spur, into bright, open canopy once again. For the first time we see the mountain's vastness rising powerfully above us. We stand for a moment and stare, saying nothing.

The foliage is sopping wet. Miniature white strings of orchid pearls (*Pholidota pectinata*) dangle from the drippy, moss-covered branches. We encounter the occasional creeping vine of *N. villosa* along the way but most don't have pitchers. But 4.5 km (2.8 miles) into the trek I find a superb one! It's growing out of a ferny escarpment about 2 metres above my head and has about 20 large pitchers cascading over the rock. Stout woody ones jostle among fresh apple-green and red ones, all suspended at different heights. Away from the track I find yet more pitchers, half-swallowed up by moss. Their yellow and red mouths yawn widely. I make sketches, hastily, mindful of the climb that still lies ahead. We press on.

2,700 m (8,858 ft). Just above Layang-Layang ('Place of Swallows'), the path leaves the dense forest and fuses with the summit trail. Our progress up the slab-like steps slows as the air turns cooler and thinner, and the earth turns into wet, mustardy clay. Having had the mountain to ourselves all day, Ramin and I join other climbers clad with poles and red, blue and grey waterproofs. We rise haltingly up the path. Weaving through a tangle of mossy branches, the familiar Laban Rata flickers into view by 4.30 pm. I rest there for the remainder of the day and write my journal. Through the window, I watch a white blanket of mist roll in like the tide and envelop the resthouse.

Along with everybody else, I surface at 2 am. I hunt for Ramin in the bleary-eyed, menthol-soaked throng under the electric light. There's an air of collective determination in the room, like runners

Aeridostachya robusta on the descent from the summit of Mount Kinabalu to Mesilau, Borneo.

huddling at the start of a race. An hour late, an unapologetic Ramin and I set off into the gloom, donned heavily with head torches and rustling Gore-tex. Although cold and windy, mercifully there's no rain this time, so after just 10 minutes' rest at the Sayat Sayat checkpoint, we haul our way up to Low's Peak. We wrestle through the 'danger zone' along guide ropes and slither over large black boulders. Desperately cold and dripping-wet from the mist, we spend all but three minutes on the bleak and perilous peak, which we've done well to reach in under two hours.

We hobble and slide back to the Laban Rata, admiring sensational views of Sabah – mile upon mile of blue-green mile, spooling out in front of us. The rocky slopes behind us become backlit with pools of orange sunlight. I can see why you would climb the mountain now, even if you were *not* obsessed with plants. The descent to Mesilau is far harder than the climb up. My limbs ache and our pace slackens as the trail dips in and out of forested hills and valleys. Fortunately the effort is more than compensated for by a stand of *Aeridostachya robusta* poking out of a deceased tree. The chunky orchid is shooting out thick, arching sprays of pinkish flowers into the gloom. I run out of water, so we replenish our bottles from a cold stream. We emerge from the dark, thicketed path into a sunny opening and I realise, to my relief, that we have reached Mesilau. I thank Ramin for all his help in searching for plants and reaching the summit. He tells me it has been his pleasure, and then asks for a RM20 tip.

IT'S A PITY about the golf course: that one of the most floristically diverse corners of our biosphere was cleared to grow grass. Plants and animals shaped over hundreds of millions of years, wiped out in a heartbeat. I suppose it was put there because of the views, at a time when we all thought nature could just move to one side. Before we even knew about things such as climate change, melting glaciers, sea level changes, the plastic. The scale of destruction. But I can't be weighed down by that now, not today. I head for the old landslip where one of the largest populations of king pitcher plants (*Nepenthes rajah*) still grows. The little trail begins in mountain rainforest, with scarlet ginger flowers gawking at me by a rickety bridge spanning a splashy stream. Nearby, a small '*Nepenthes* garden' has been planted for conservation purposes. It contains an assortment of natural hybrids amassed from around the Kinabalu Park. Many are new to me, and I spend a golden minute admiring a gallery of undreamed-of pitchers.

As I make my way over the nose of the mountain, I hear the Mesilau East river gushing furiously beneath me to my right. Above me, a steep slope is covered in small trees with black-wet, pole-like trunks splashed with lichens, arching ferns and sedges and the occasional mountain orchid. The air smells wet, like morning dew. I make my way slowly, still weary from the climb. But fatigue vanishes at the sight of dozens of giant burgundy pitchers with gaping yellow mouths. They are everywhere – hanging from branches, festooning the little trees, and draping over the wet, rocky, orange and green ledges. Yes, I'm surrounded by them! Enormous matte grey-green leaves are thrusting out all over the place, their tendrils groping for places to send their developing pitchers. Some have done so right onto the track. In a hummock of wet, grey scrub, I find the finest one. It's an impossibly large specimen with an oval scooped lid the size of a dinner plate, and a glossy, yellow-green interior. A king pitcher on its rocky throne. I peer inside its bucket-like structure. It's grotesque and beautiful at the same time – don't ask me how, but it is. I watch the brownish syrupy pond slosh

from side to side as I hold it. I see mosquito larvae, like hyphens twitching to the bottom of the pool. To sit with giant pitcher plants on a misty mountainside in Borneo is an unforgettable experience. One that doesn't seem to belong to this world.

> *Another steep climb of 800 feet brought us to the Marei Parei spur, to the spot where the ground was covered with the magnificent pitcher-plants, of which we had come in search. This one has been called the* Nepenthes Rajah, *and it is a plant about four feet in length, with broad leaves stretching on every side, having the great pitchers resting on the ground in a circle around it. Their shape and size are remarkable . . . It is indeed one of the most astonishing productions of nature.*
> Spenser St. John, 1858.[55]

Tragically, on 5 June 2015 Mount Kinabalu was struck by an earthquake of a magnitude not seen since the 1970s. Eighteen people, including hikers and mountain guides, were killed by the earthquake and the massive landslide that followed. Mesilau was affected particularly badly, and much of its trail – now closed indefinitely – and the wonderful plants that grew along it exist no more.

Nepenthes rajah growing in the hills above the Mesilau East river, Mount Kinabalu, Borneo.

I SAW A ghost today. If you saw it too, you'd know what I mean. *Nepenthes burbidgae* is a spectre of a pitcher plant, like a negative print. Its ivory-cream pitchers are splashed with a shade of magenta artists dream of. Known as the 'painted pitcher plant', it grows at about 2,000 m (6,561 ft) elevation on ultramafic (igneous) slopes of the Pinosok Plateau along with *N. rajah* and their handsome hybrid *N.* x *alisaputrana* (*N. rajah* x *N. burbidgeae*). Nothing can match the monstrous proportions of the pitchers of *N. rajah* I saw yesterday; still, the painted pitcher plant has its own ethereal beauty. Its porcelain pitchers glow against the black, mossy earth. Lanterns. Yes, that's what they remind me of.

The species was discovered by Hugh Low and Spenser St. John in 1858, and named after Frederick William Burbidge, an explorer who collected tropical plants for the famous Veitch Nurseries and made his own ascent of the mountain in 1877. I'm reminded of their encounters with the plant one and a half centuries before my own:

> *Crossing the Hobang, a steep climb led us to the western*
> *spur, along which our path lay; here, at about 4000 feet,*
> *Mr. Low found a beautiful white and spotted pitcher-plant*
> *which he considered the prettiest of the twenty-two species*
> *of nepenthes with which he was then acquainted: the pitchers*
> *are white and covered in a most beautiful manner with spots*
> *of an irregular form, of a rosy pink colour . . . it is a climbing*
> *plant, and varies from fifteen to twenty feet in length.*
> Spenser St. John, 1862.[56]

Nepenthes burbidgae, on Mount Kinabalu's Pinosok Plateau, Borneo.

THE MOUNTAIN GARDEN is nestled in a forest at the end of an enclosed, shaded track. The air is pleasantly sweet and refreshing here, about 520 m (1,706 ft) above sea level, and the cold Silau-Silau mountain stream knocks and gurgles its way through the lushness. Neglected by tourists, this oasis contains some of Kinabalu's rarest botanical treasure. It's the only place to see so many of its plants in proximity (and carries a lower risk of falling down a precipice). The quiet, laissez-faire garden is crammed full of rare orchids and pitcher plants snuggling the trunks of emerald leafy shrubs, floppy wet banana leaves and bunches of plastic pink *Medinilla* fruits. I can't see anyone working here, but whoever does must love the place.

I find two specimens of *N. rajah,* one without pitchers, another with modest-sized, crimson-marbled ones. They're far less impressive than the wild populations, possibly because of the shadier conditions. At the rear of the garden I find a fenced-off area where plants grow under lock and key. Among the shelves packed with plants escaping from their little pots, I see Rothschild's slipper orchid (*Paphiopedilum rothschildianum*). This is the State Flower of Sabah and worthy of its title, the 'Gold of Kinabalu'. Several flowers are open, each with polka-dotted sepals that spread-eagle to the width of a dinner plate. Nearby, I see 'sexy lady orchids' (*Stichorchis*) equally worthy of *their* title. I laugh when I see the outline of a buxom female form in their tiny green flowers.

Most trees in the garden are veiled in moss and some have enormous bird's-nest ferns sprouting out of their forked trunks. One tree grown here is a form of *Tristaniopsis*, with an orange and blue-grey trunk as smooth as marble. It has the feel of a eucalyptus (to which it's distantly related). By violent contrast, the stem of a nearby rattan (*Plectocomia elongata*) is snared with

Nepenthes x *alisaputrana*, on Mount Kinabalu's Pinosok Plateau, Borneo.

ranks of vicious thorns, capable of inflicting serious damage. Ants march up and down it purposefully.

In a forgotten corner of the garden I find a rare hybrid N. *lowii* x N. *stenophylla* with large, jar-like pitchers. About a metre away I find its parent, N. *lowii,* with extraordinary chalices hanging down from the branches. A liana-like vine of N. *edwardsiana* is throwing out a handful of orangey-red pitchers, with contrasting interiors the colour of apple flesh. For its size, this single hectare of garden might have one of the richest assemblages of plants in the world.

Quenched with pitcher plants, I make my way back to headquarters for the bus to Kota Kinabalu. Sharp notes of rotting fruit and vomit waft through the bus. But I feel calm and content as I watch the world's greatest botanical monument fade from view. I've a strange, stirring feeling inside me, as if something important has just happened but I'm not sure what.

EVERY ADDICTION TAKES its toll. On your body, on your mind. On your soul. The one that bought me here, the one that took me slithering and scrambling my slippery way up rocks and down ropes, dragged me up trees and onto outcrops, has taken its toll. Today my legs hurt so badly I could barely get out of bed. A month spent trekking up and down mountains will do that to you. But you will stop at nothing to get that next fix – that's how addictions work. I always need to find something wilder, more remarkable. Something big. Listen, I will not rest until I've found it. Because once you've divined the world's largest flower in your mind's eye, you have to do it for real, don't you? Find it, whatever it takes. Get to know its haunts; bathe in its beauty.

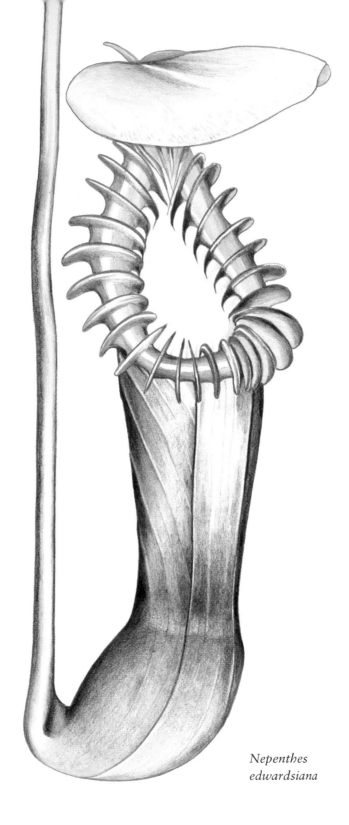

Nepenthes
edwardsiana

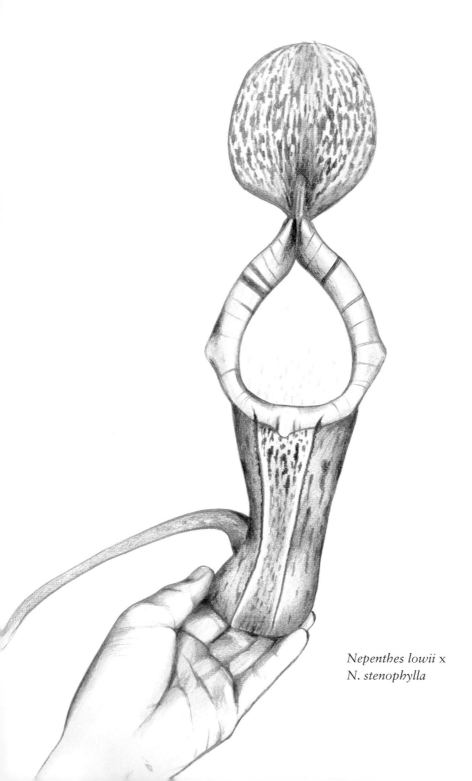

Nepenthes lowii x
N. stenophylla

OUR DESTINATION TODAY is the rainforest around Poring, an area at the foot of the mountain, famous for its hot springs. Here we plan to see the low elevation forest flora, and hunt for a giant. I'm joined by two tourists, a Japanese woman and an Italian man, along with our talkative guide who points out interesting trees along the way. We stop at a market not far from the mountain and sample 'tarap' (*Artocarpus odoratissimus*) – another large, foetid fruit (these do seem popular here!). Seven of the olive brown orbs are strung up – one more and I suspect they would bring the whole shelter down. Rather like a jackfruit but far superior, I'm fervently assured; once prised open, a collection of strange white, slimy structures like sea squirts spew out. It tastes like warm vanilla ice cream, were such a thing to exist. We also sample lychee-like rambutans (*Nephelium lappaceum*) and various other delicacies. People are very proud of the fruits that grow here. I think it best not to mention that what I truly long for is a banana.

The rainforest is dominated by ranks of towering dipterocarp trees. The pulsing trill of insects is intense and tropical birds echo about the trees. As we penetrate the hissing wilderness we see all manner of tropical crops growing exuberantly out of the warm wetness, including figs, gingers, yams, tarap trees and planted tapioca plants (*Manihot esculenta*) with cannabis-like fronds. Enormous stands of bamboo and giant elephant ears (*Alocasia macrorrhizos*) rise through the gloom. The impossibly large leaves of the elephant ear are used as umbrellas during typhoons. I could do with one of those most afternoons out here. Arrow-straight trees, blackened by the rain, lead my eye up into the canopy where white light is intercepted by the silhouettes of a million leafy fragments; loops of liana lead it back down again. Suddenly, in the green chaos, we meet with something astonishing: a fruiting spike of a giant *Amorphophallus lambii* has erupted out of the forest floor. Its fat stem holds a cylinder of glassy orange berries that glow against the leaf litter as if they are lit from within. After some searching we find a flowering specimen. Now we're talking. I hop up and down with delight next to the plant while the others look on and laugh. Its spathe (the large, cowl-like structure) is leathery, wrinkled and maroon, and it has a fat, grey-blue spadix (this is the spike where the flowers are produced, at the base). It's a relative of the titan arum (*A. titanum*) – the one that makes the headlines every time it flowers in botanic gardens. I remember being taken to see one at Kew Gardens as a child; remember staring up at it, feeling something seethe inside me. Dopamine, probably.

After lunch, the others make their way to a natural spring in the forest, but I remain fixed to the spot, taking a precious moment to absorb it all. Because I'm entranced by this place; completely under its green spell. Hundreds of plants enclose me. I peer at the unfamiliar ones, turning over this leaf and that. They could be rare and endangered; new to science; harbour cures for illnesses that don't even exist yet. I don't know. I have my whole life ahead of me to dedicate to them. But I'm young and impatient. And there's a plant I need to see now.

'Come with me, Chris, and I show you the special flower,' whispers my guide, excited and grinning. We leave the others at the spring and pick our way through rainforest clearings among herds of cows, then clamber down a slope until we reach a shady, forested hill fringed with ranks of bamboo. The

Amorphophallus lambii, in the rainforest at Poring, Borneo.

landowner has protected this small area specifically for income from eco-tourists eager to tick off seeing the largest flower in the world. As we advance, the air becomes charged with the electric hiss of a million insects. They scream at me to dive in. We enter the forest, whispering, as if the plants can hear our approach.

Then, as if in a dream, I'm standing with the pack. Corpse flowers (*Rafflesia keithii*). My heart quickens. There are several, each at various stages of development. One unfurled just a week ago and is fully open. I'm drawn to it like a magnet, but slowly, as if I'm moving underwater. It's otherworldly. Just like the pictures in the cherished books of my childhood; just like the pictures in my mind; the premonitions, the dreams, yes, all of it, coming to life. I kneel by it. Peer into it. Absorb its colours. The flower is about a metre across, and has warty petals the colour of rust, draping onto the floor. At its centre, a domed chamber conceals a spiny disc; it looks like an intricate Italian meringue, complete with little brown turrets. Then, around the flower, I notice buds too, sprouting out of the warm, wet earth. They are like pink cabbages. One is the size of a football, and it looks as if it might burst open right in front of me. Smaller ones too, the size of tennis balls, are growing in lines, and these have been carefully concealed with leaves by the landowner to deter unsought interest. On a slope nearby, we see the rotting remains of spent flowers, all black and slimy, and small, crater-like scars on the *Tetrastigma* vines from which they are feeding. I can see the whole process unfolding all around me. I examine a bud and stroke it gently. It looks full of promise. I wonder who'll marvel at this one when the time comes, I whisper to myself.

When it's time to leave, I wipe away the sweat from my cheek and realise, embarrassed, that I've shed a tear. Probably all the exhilaration – like the kid who opened all his Christmas presents at once – that sort of thing. Carefully, we rearrange the leaf litter and quietly pick our way out of the clearing, crossing a little bamboo bridge over a stream and leaving the place just as we found it.

The first British man to see a Rafflesia *was the naturalist Dr Joseph Arnold in 1818, while on expedition to Western Sumatra. A Malay servant (unnamed in Arnold's account) led him to a specimen with the words, 'Come with me, Sir, come! A flower, very large, very beautiful, wonderful!' Arnold's written account of the experience begins, 'I rejoice to tell you that I happened to meet with what I consider as the greatest prodigy of the vegetable world.'*[57]

NOW THAT WE'RE back I'll tell you this: it is not enough. Not enough that I've dreamed of them, spent time with them, become addicted to and possessed by them, chased them down cliffs, fallen for them and stared at the sun. No. You can't hold on to that amount of electricity and do nothing with it, can you? You'd explode. Here's what I do: I lie awake in the half-light and I let it ground. I watch the plants slowly unfurl across the ceiling – the colours, the shapes – drawing strength, coming to life, taking over. Sometimes I watch them die there too. Because I've learned that on canvas what brings a plant to life is death – a withered leaf, a nibble, a wrinkle. Yes, patiently I wait for their stems to spread above me, creep down the walls, weave around one another, interlock and find their place. Then come the details. *Details, details.* Every hair, vein and grain of pollen – all of it.

And then I paint.

Rafflesia keithii in the
rainforest at Poring, Borneo.

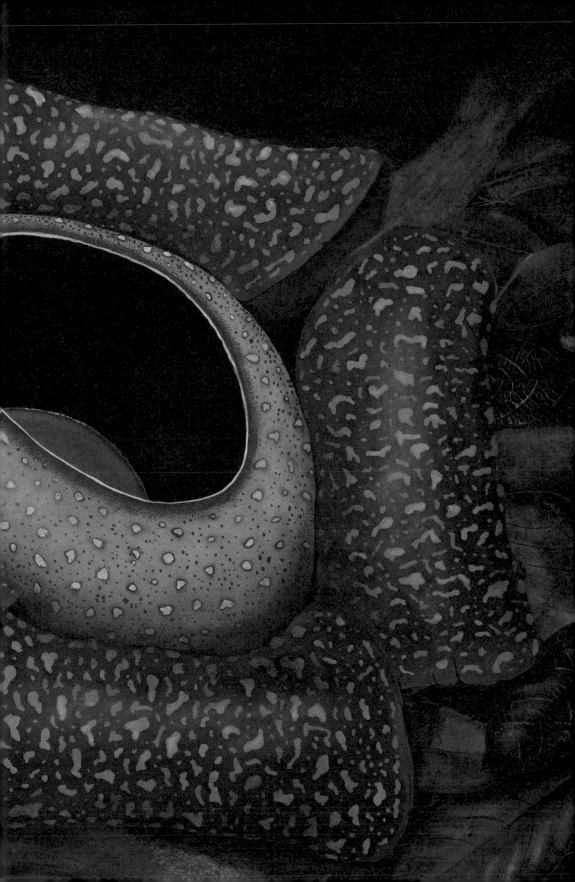

ACKNOWLEDGEMENTS

LIFE AS A BOTANIST has turned out to be a scream. Many of the people who join me on these pages made it this way. Thank you, in order of appearance to: my loved ones for giving me the freedom to be the toad-training, broomrape-hunting, clifftop-dangling botanist I am; William Bortrick for his hospitality in Northern Ireland; Fred Rumsey for his patience (everywhere). Thanks to the helpful staff at the Karoo Desert National Botanical Garden in South Africa; and in Israel to my friends Yuval Sapir (Director, the Botanical Garden of Tel Aviv), Ori Fragman-Sapir (Scientific Director of the Jerusalem Botanical Gardens) and botanist Dar Ben-Natan. Thanks to Alfredo Reyes-Betancort (Director of Jardín de Aclimatación de La Orotava) for showing me every plant species on Lanzarote, and to Matías Hernandez Gonzalez (Founder of Arrecife Natura), friends, and family, for making me feel so at home on their beautiful island. I owe a debt of thanks to many people for the excursions in Japan: Ben Jones (for putting up with me!); Alison Beale for her hospitality on location, and many helpful suggestions to improve chapter 6, *Across Kingdoms*. My sincere thanks to the many Japanese botanists who tirelessly shared their flora with us, one prefecture at a time: Nakata-san (Director), Godo-san, Ohara-san and the rest of the team at the Toyama Botanic Gardens; Sakaue-san and Kimura-san of the University of Tokyo; Ayako-san and Michiyo-san of Makino Botanical Garden; and Atsushi-san and Kensei-san of the Okinawa Churashima Foundation. Thanks also to the mountain guides of Malaysian Borneo for sharing their extraordinary flora with me so patiently. Thank you also to the many characters around the world, whose names I'll never know, with whom I have spoken the language of plants.

The trips to Japan would not have been possible without the generosity of Jane Impey and Junko Oikawa, who helped establish the relationship with Oxford. Excursions to the Canary Islands and the Middle East were sponsored by Linacre College.

Finally I'd like to thank Lydia White and Gina Fullerlove of the Kew Publishing team, and Joseph Calamia of the University of Chicago Press team. No author could wish for a better team to work with.

NOTES

1   C. J. Thorogood U. Bauer and S. J. Hiscock, 'Convergent and Divergent Evolution in Carnivorous Pitcher Plant Traps', *New Phytologist* 217 (2017): 1035–41.

2   F. Box, C. J. Thorogood and J. Hui Guan, 'Guided Droplet Transport on Synthetic Slippery Surfaces Inspired by a Pitcher Plant', *J. R. Soc. Interface* 16 (158) (September 2019): 20190323.

3   C. J. Thorogood, C. J. Leon, D. Lei, M. Aldughayman, L-f Huang and J. A. Hawkins, 'Desert Hyacinths: An Obscure Solution to a Global Problem?' *Plants, People, Planet* 3 (2021): 302–7.

4   M. Y. Siti-Munirah, N. Dome and C. J. Thorogood, '*Thismia sitimeriamiae* (Thismiaceae), an extraordinary new species from Terengganu, Peninsular Malaysia', *PhytoKeys* 179 (2021): 75–89.

5   A. Antonelli, C. Fry, R. J. Smith, M. S. J. Simmonds, P. J. Kersey et al., *State of the World's Plants and Fungi 2020*. Royal Botanic Gardens, Kew.

6   A. Pratt, *The Flowering Plants and Ferns of Great Britain*. London: The Society for Promoting Christian Knowledge, 1855.

7   S. A. Harris, *The Magnificent Flora Graeca: How the Mediterranean Came to the English Garden*. Oxford: Bodleian Publishing, 2007.

8   C. A. Thanos, 'Aristotle and Theophrastus on Plant-Animal Interactions', in *Plant-Animal Interactions in Mediterranean-Type Ecosystems*, ed. M Arianoutsou and R. H. Groves. Dordrecht: Springer, 1994.

9   L. F. Haas, 'Pedanius Dioscorides (born about AD 40, died about AD 90)', *Journal of Neurology, Neurosurgery and Psychiatry* 60, 4 (1996): 427.

10 J. Sibthorp, in *Memoirs relating to European and Asiatic Turkey and other Countries of the East edited from the manuscript journals*, ed. R. Walpole (1818). London: Longman, Hurst, Rees, Orme and Brown, 1975, 66.

11 J. Sibthorp, in *Memoirs relating to European and Asiatic Turkey and other Countries of the East edited from the manuscript journals*, ed. R. Walpole (1818). London: Longman, Hurst, Rees, Orme and Brown, 1974, 89.

[12] C. R. Darwin, *On the Various Contrivances by which British and Foreign Orchids are Fertilised by Insects, and on the Good Effects of Intercrossing*. London: John Murray, 1862.

[13] E. Coleman, 'Pollination of the Orchid *Cryptostylis leptochila*, *Victorian Naturalist* 44 (1927): 20–2.

[14] J. Sibthorp, in *Memoirs relating to European and Asiatic Turkey and other Countries of the East edited from the manuscript journals*, ed. R. Walpole (1820). London: Longman, Hurst, Rees, Orme and Brown, 1975, 82.

[15] J. Sibthorp, in *Memoirs relating to European and Asiatic Turkey and other Countries of the East edited from the manuscript journals*, ed. R. Walpole (1820). London: Longman, Hurst, Rees, Orme and Brown, 1987, 32.

[16] J. Sibthorp, in *Memoirs relating to European and Asiatic Turkey and other Countries of the East edited from the manuscript journals*, ed. R. Walpole (1820). London: Longman, Hurst, Rees, Orme and Brown, 1987, 37.

[17] G. Durrell, *My Family and Other Animals*. London: Penguin Books, 1956.

[18] J. Sibthorp, in *Memoirs relating to European and Asiatic Turkey and other Countries of the East edited from the manuscript journals*, ed. R. Walpole (1820). London: Longman, Hurst, Rees, Orme and Brown, 1974, 18.

[19] J. Sibthorp, in *Memoirs relating to European and Asiatic Turkey and other Countries of the East edited from the manuscript journals*, ed. R. Walpole (1820). London: Longman, Hurst, Rees, Orme and Brown, 1987, 27–8.

[20] S. Lieberman, *Sumerian Loanwords in Old Babylonian Akkadian*. Missoula: Scholars Press, 1977, 475.

[21] H. Adams, '1907', in *The Education of Henry Adams*. ed. H. Adams. New York: Modern Library, 1931.

[22] T. Shaw, *Travels or Observations Relating to Several Parts of Barbary and the Levant*, Vol. II. Edinburgh: J. Ritchie, 1808.

[23] A. V. Humboldt and A. Bonpland, *Personal Narrative of Travels to the Equinoctial Regions of the New Continent During the Years 1799–1804* (1826), translated by H. M. Williams. New York: Cambridge University Press, 2011, 274.

24 A. V. Humboldt and A. Bonpland, *Personal Narrative of Travels to the Equinoctial Regions of the New Continent During the Years 1799–1804* (1826), translated by H. M. Williams. New York: Cambridge University Press, 2011, 83.

25 A. V. Humboldt and A. Bonpland, *Personal Narrative of Travels to the Equinoctial Regions of the New Continent During the Years 1799–1804* (1826), translated by H. M. Williams. New York: Cambridge University Press, 2011, 80.

26 A. V. Humboldt and A. Bonpland, *Personal Narrative of Travels to the Equinoctial Regions of the New Continent During the Years 1799–1804* (1826), translated by H. M. Williams. New York: Cambridge University Press, 2011, 95.

27 A. V. Humboldt and A. Bonpland, *Personal Narrative of Travels to the Equinoctial Regions of the New Continent During the Years 1799–1804* (1826), translated by H. M. Williams. New York: Cambridge University Press, 2011, 159.

28 A. V. Humboldt and A. Bonpland, *Personal Narrative of Travels to the Equinoctial Regions of the New Continent During the Years 1799–1804* (1826), translated by H. M. Williams. New York: Cambridge University Press, 2011, 90.

29 C. P. Thunberg, *Travels in Europe, Africa, and Asia, Performed Between the Years 1770 and 1779*, Vol. III containing 'A Voyage to Japan and Travels in Different Parts of that Empire in the years 1775 and 1776'. London: printed for F. and C. Rivington, 1796.

30 K. J. W. Hensen, 'Identification of the Hostas ("Funkias") Introduced and Cultivated by Von Siebold', *Mededelingen Van de Landbouwhogeschool te Wageningen, Nederland* 63, 6 (1963): 1–22.

31 A. Le Lievre, 'Carl Johann Maximowicz (1827–91), Explorer and Plant Collector', *The New Plantsman* 4, 3 (1997): 131–43.

32 W. Weston, *The Playground of the Far East*. London: John Murray, 1918.

33 W. Weston, *The Playground of the Far East*. London: John Murray, 1918, 78.

34 C. P. Thunberg, *Travels in Europe, Africa, and Asia, Performed Between the Years 1770 and 1779*, Vol III containing 'A Voyage to Japan and Travels in Different Parts of that Empire in the years 1775 and 1776'. London: printed for F. and C. Rivington, 1796, 214.

35 Matsuo Bashō, *Narrow Road to the Interior and Other Writings* (1644–94), translated by S. Hamill. Boston: Shambhala Publications, 1998.

36 J. Fisher, *Wild Flowers in Danger*. London: H. F. & G. Witherbey LTD, 1987, 61.

37 Matsuo Bashō, *The Narrow Road to the Deep North and Other Travel Sketches*, translated by Nobuyuki Yuasa. London: Penguin Books, 2005.

38 C. P. Thunberg, *Travels in Europe, Africa, and Asia, Performed Between the Years 1770 and 1779*, Vol. III containing 'A Voyage to Japan and Travels in Different Parts of that Empire in the years 1775 and 1776'. London: printed for F. and C. Rivington, 1796, 164–5.

39 W. Weston, *The Playground of the Far East*. London: John Murray, 1918, 200.

40 W. Weston, *The Playground of the Far East*. London: John Murray, 1918, 128.

41 C. P. Thunberg, *Travels in Europe, Africa, and Asia, Performed Between the Years 1770 and 1779*, Vol. III containing 'A Voyage to Japan and Travels in Different Parts of that Empire in the years 1775 and 1776'. London: printed for F. and C. Rivington, 1796, 227.

42 Matsuo Bashō, *Bashō's Haiku: Selected Poems of Matsuo Bashō*, translated by David Landis Barnhill. New York: University of New York Press, 2004, 21.

43 W. Weston, *The Playground of the Far East*. London: John Murray, 1918, 205.

44 C. P. Thunberg, *Travels in Europe, Africa, and Asia, Performed Between the Years 1770 and 1779*, Vol. III containing 'A Voyage to Japan and Travels in Different Parts of that Empire in the years 1775 and 1776'. London: printed for F. and C. Rivington, 1796, 83.

45 W. Weston, *The Playground of the Far East*. London: John Murray, 1918, 142.

46 W. Weston, *The Playground of the Far East*. London: John Murray, 1918, 143.

47 W. Weston, *The Playground of the Far East*. London: John Murray, 1918, 105.

48 W. Weston, *The Playground of the Far East*. London: John Murray, 1918, 217.

49 C. P. Thunberg, *Travels in Europe, Africa, and Asia, Performed Between the Years 1770 and 1779*, Vol. III containing 'A Voyage to Japan and Travels in Different Parts of that Empire in the years 1775 and 1776'. London: printed for F. and C. Rivington, 1796, 8.

50 Matsuo Bashō, *Bashō's Journey: The Literary Prose of Matsuo Bashō*, translated by David Landis Barnhill. New York: University of New York Press, 2005, 4.

51 L. S. Gibbs, 'Contribution to the Flora and Plant Formations of Mount Kinabalu and the Highlands of British North Borneo', *Journal of the Linnean Society*, Vol. XLII, 1913.

52 E. J. H. Corner, 'The Plant Life', in *Kinabalu: Summit of Borneo,* (Kota Kinabalu: Sabah Society, 1978).

53 J. Hooker, 'On the Origin and Development of the Pitchers of *Nepenthes*, with an Account of some New Bornean Plants of that Genus. *Transactions of the Linnaean Society* 22 (1859): 415–21.

54 F. W. Burbidge, *The Gardens of the Sun*. London: John Murray, 1880 100.

55 S. St. John, *Life in the Forests of the Far East*, Vol. 1. London: Oxford University Press, 1862, 324.

56 S. St. John, *Life in the Forests of the Far East*, Vol. 1. London: Oxford University Press, 1862, 323.

57 R. Brown, *An Account of a New Genus of Plants, Named* Rafflesia. London: printed by Richard and Arthur Taylor, 1821, 2.

Pitcher plants (*Nepenthes villosa*)
growing on Mount Kinabalu in Borneo.

pp 276–7

It took me 16 years to start this painting. I can still remember exactly where I was standing when it was conceived: can smell the moss, feel the wet velvet pitchers under my fingertips. Painting transports me back to a place in a way that is hard to explain. I need to be in the painting: hyperrealism – a style of illustration that resembles high-definition photography – is what takes me there.

I start by sketching on scraps of paper, then I scale up. I work at life size; for these plants, that means A0 (841 x 1,188 mm or 33 x 47 inches). Once I'm satisfied with the composition, I fill the empty shapes with primer. I'm not fussy – wall paint will do. Then, in oils, I start the background. I paint in thin layers (glazes), which build depth. Many of the details get lost along the way. Those misty hills on the left – they had intricate branches that got lost in a vale of cloud. This may sound like a waste of time, but it's important for me to know these details are there, even if you can't see them.

I use a dozen brushes, but most of what you see is painted with a rigger brush. Riggers have long hairs that absorb the shakes – people once used them to depict the rigging of boats. Slow, steady sweeps – that's the trick; any hesitation will be captured in paint, and I can't allow that. The fumes are heady, but I get too absorbed in detail to notice. After the hills come the mosses; after the mosses come the orchids – then there's no stopping them: steadily the plants take over. The pitcher plants come last. Painting in sequence like this brings subjects in or out of focus; it allows you to reach out and touch the plants at the front.

Every night I stare at it, while cleaning my teeth, and I see imperfections; things I need to put right. I lay awake taunted by them. It's painstakingly slow too. A painting this size takes me about two months to complete. Two months and 172,800 brushstrokes. That's how long it takes to fall out of love with a painting. Once I no longer want to look at it: that's when I know it's done.

INDEX OF PLANTS

Images are indicated by page numbers in **bold**.

INDEX OF PLACES

Images are indicated by page numbers in **bold**.